AN INTRODUCTION TO ALGORITHMIC TRADING

Basic to Advanced Strategies

Edward A Leshik
Jane Cralle

WILEY
A John Wiley and Sons, Ltd., Publication

This edition first published 2011.

Registered office
John Wiley & Sons Ltd, The Atrium, Southern Gate, Chichester, West Sussex, PO19 8SQ, United Kingdom

For details of our global editorial offices, for customer services and for information about how to apply for permission to reuse the copyright material in this book please see our website at www.wiley.com.

Reprinted May 2011

For other titles in the Wiley Trading Series
Please see www.wiley.com/finance

A catalogue record for this book is available from the British Library

ISBN 978-0-470-68954-7 (hardback); ISBN 978-0-470-97935-8 (ebk);
ISBN 978-1-119-97509-0 (ebk); ISBN 978-1-119-97510-6 (ebk)

Typeset in 10/12pt Times by Aptara Inc., New Delhi, India
Printed in Great Britain by TJ International Ltd, Padstow, Cornwall

Contents

PART II THE LESHIK-CRALLE TRADING METHODS

Acknowledgments

Edward

Thanks go to Gerry Prosser, Rob Bruce, Ian Kaplan, Dr. Mivart Thomas, Sebastian Thomas, Jason Sharland, Dan Fultz and the late Gillian Ferguson. To dearest Diana go thanks for fifty years of enthusiasm, encouragement, wisdom and insight – truly a 'woman for all seasons'.

EDWARD LESHIK
London, England

My acknowledgements from the western world:

Jane

Could not have done it without you folks –

Lisa Cralle Foster, J. Richard (Rick) Kremer, FAIA, Alan H. Donhoff, Lisa Luckett Cooper, Rose Davis Smith, Helen D. Joseph, Shelly Gerber Tomaszewski, Brad Kremer, Jenny Scott Kremer and the late John Ed Pearce. Then there is Mr. Linker, President of Linker Capital Management Inc., an honor to his father, the late Samuel Harry Linker.

JANE CRALLE
Kentucky, USA

Both Edward and Jane

Our sincere thanks go to Aimee Dibbens for her encouragement and enthusiasm in getting this book written. Special thanks to the great team at Wiley, Peter Baker, Vivienne Wickham, Caroline Valia-Kollery, Felicity Watts and the anonymous reviewers who helped shape this book.

Our special thanks go to Nick Atar whose enthusiastic encouragement and hospitality at the Waffle helped make this book a reality.

Mission Statement

The goal of this book is to:

1. Demystify algorithmic trading, provide some background on the state of the art, and explain who the major players are.
2. Provide brief descriptions of current algorithmic strategies and their user properties.
3. Provide some templates and tools for the individual trader to be able to learn a number of our proprietary strategies to take up-to-date control over his trading, thus level the playing field and at the same time provide a flavor of algorithmic trading.
4. Outline the math and statistics we have used in the book while keeping the math content to a minimum.
5. Provide the requisite Excel information and explanations of formulas and functions to be able to handle the algorithms on the CD.
6. Provide the reader with an outline 'grid' of the algorithmic trading business so that further knowledge and experience can be 'slotted' into this grid.
7. Use a 'first principles' approach to the strategies for algorithmic trading to provide the necessary bedrock on which to build from basic to advanced strategies.
8. Describe the proprietary ALPHA ALGOS in Part II of the book to provide a solid foundation for later running of fully automated systems.
9. Make the book as self-contained as possible to improve convenience of use and reduce the time to get up and running.
10. Touch upon relevant disciplines which may be helpful in understanding the underlying principles involved in the strategy of designing and using trading algorithms.
11. Provide a detailed view of some of our Watchlist of stocks, with descriptions of each company's operations. Provide a framework for analyzing each company's trading characteristics using our proprietary metrics. It is our belief that an intimate knowledge of each stock that is traded provides a competitive advantage to the individual trader enabling a better choice and implementation of algo strategies.

Part I
INTRODUCTION TO TRADING ALGORITHMS

Preface to Part I

Fabrizzio hit the SNOOZE he was dreaming he hit the TRADE key and within 15 milliseconds hundreds of algorithms whirred into life to begin working his carefully prethought commands. ALARM went off again, time to get up with the haze of the dream session End of day lingering, net for the day $10 000 000 . . . not bad, not bad at all, he smiled as he went into his 'getting to work routine.'

Can we trade like that? Answering this question is what this book is all about.

Algorithmic trading has taken the financial world by storm. In the US equities markets algorithmic trading is now mainstream.

It is one of the fastest paradigm shifts we have seen in our involvement with the markets over the past 30 years. In addition there are a number of side developments operated by the Tier 1 corporations which are currently the subject of much controversy and discussion – these are based, to a great extent, on 'controversial' practices available only to the Tier 1 players who can deploy massive resources which disadvantage the individual, resource-limited, market participant.

No doubt the regulatory machinery will find a suitable compromise in the near future and perhaps curtail some of the more flagrant breaches of ethics and fair play – an area in which Wall Street has rarely excelled and now could well do with some help to restore the dented confidence of the mass public.

Notwithstanding these side issues, the explosive growth of algorithmic trading is a fact, and here to stay.

Let us examine some of the possible reasons for such a major and dramatic shift.

We believe the main reasons for this explosive growth of algorithmic trading are: Rapid cost reduction; better controls; reduction of market impact cost; higher probability of successful trade execution; speed, anonymity and secrecy all being pushed hard by market growth; globalization and the increase in competition; and the huge strides in advancing sophisticated and available technology.

In addition there is also the conceptual and huge advantage in executing these carefully 'prethought' strategies at warp speed using computer automation all of which would be well beyond the physical capability of a human trader.

Algorithmic trading offers many advantages besides the ability to 'prethink' a strategy. The human trader is spared the real-time emotional involvement with the trade, one of the main sources of 'burn out' in young talented traders. So in the medium term there is a manpower saving which, however, may be offset by the requirement for a different type of employee with more expensive qualifications and training.

Algos can execute complex math in real time and take the required decisions based on the strategy defined without human intervention and send the trade for execution automatically from the computer to the Exchange. We are no longer limited by human 'bandwidth.' A computer can easily trade hundreds of issues simultaneously using advanced algorithms with layers of conditional rules. This capability on its own would be enough to power the growth of algorithmic trading due to cost savings alone.

As the developments in computer technology facilitated the real-time analysis of price movement combined with the introduction of various other technologies, this all culminated in algorithmic trading becoming an absolute must for survival – both for the Buy side and the Sell side and in fact any serious major trader has had to migrate to the use of automated algorithmic trading in order to stay competitive.

A Citigroup report estimates that well over 50% of all USA equity trades are currently handled algorithmically by computers with no or minimal human trader intervention (mid-2009). There is considerable disagreement in the statistics from other sources and the number of automated algorithmic trades may be considerably higher. A figure of 75% is quoted by one of the major US banks. Due to the secrecy so prevalent in this industry it is not really possible to do better than take an informed guess.

On the cost advantage of the most basic automated algorithmic trading alone (estimated roughly at 6 cents per share manual, 1 cent per share algorithmic) this is a substantial competitive advantage which the brokerages cannot afford to ignore. Exponential growth is virtually assured over the next few years.

As the markets evolve, the recruitment and training of new algo designers is needed. They have to be constantly aware of any regulatory and systemic changes that may impact their work. A fairly high level of innate intellectual skill and a natural talent for solving algorithmic area problems is a 'must have' requirement.

This is changing the culture of both the Buy side and Sell side companies. Many traders are replaced by 'quants' and there is a strong feeling on the Street of 'physics' envy. A rather misplaced and forlorn hope that the ability to handle 3rd order differential equations will somehow magically produce a competitive trading edge, perhaps even a glimpse of the 'Holy Grail,' ALPHA on a plate.

As the perception grows in academic circles that the markets are 'multi-agent adaptive systems' in a constant state of evolution, far from equilibrium, it is quite reasonable and no longer surprising when we observe their highly complex behavior in the raw.

'Emergence,' which we loosely define as a novel and surprising development of a system which cannot be predicted from its past behavior, and 'phase transition' which is slightly more capable of concise definition as 'a precise set of conditions at which this emergent behavior occurs,' are two important concepts for the trading practitioner to understand. 'Regime' shifts in market behavior are also unpredictable from past market behavior, at least at our present state of knowledge, but the shifts are between more or less definable states.

Financial companies and governments from across the world are expected to increase their IT spending during 2010.

Findings from a study by Forrester (January 2010) predicted that global IT investment will rise by 8.1% to reach more than $1.6 trillion this year and that spending in the US will grow by 6.6% to $568 billion.

This figure may need revising upward as the flood of infrastructure vendors' marketing comes on stream.

As one often quoted Yale professor (Andrew Lo) remarked recently: 'It has become an arms race.'

Part I of this book is devoted mainly to the Tier 1 companies. We shall first describe in broad outline what algorithms are, describe some of the currently popular trading algorithms, how they are used, who uses them, their advantages and disadvantages. We also take a shot at predicting the future course of algorithmic trading.

Part II of this book is devoted to the individual trader. We shall describe the Leshik-Cralle ALPHA Algorithmic trading methodology which we have developed over a period of 12 years. This will hopefully give the individual trader some ammunition to level the trading playing field. We shall also provide a basic outline of how we design algorithms, how they work and how to apply them as an individual trader to increase your ability to secure your financial future by being in direct and personal control of your own funds.

In general we have found that successful exponents of algorithmic trading work from a wide interdisciplinary knowledge-base. We shall attempt to provide some thoughts and ideas from various disciplines we have visited along the way, if only in the briefest of outlines. Hopefully this will help to provide an 'information comfort zone' in which the individual trader can work efficiently and provide a route for deeper study.

1
History

The origin of the word 'Algorithm' can be traced to circa 820 AD when Al Kwharizmi, a Persian mathematician living in what is now Uzbekistan, wrote a 'Treatise on the Calculation with Arabic Numerals.' This was probably the foundation stone of our mathematics. He is also credited with the roots of the word 'algebra,' coming from 'al jabr' which means 'putting together.'

After a number of translations in the 12th century, the word 'algorism' morphed into our now so familiar 'algorithm.'

The word 'algorithm' and the concept are fundamental to a multitude of disciplines and provide the basis for all computation and creation of computer software.

A very short list of algorithms (we will use the familiar abbreviation 'algo' interchangeably) in use in the many disciplines would cover several pages. We shall only describe some of those which apply to implementing trading strategies.

If you are interested in algorithms per se, we recommend Steven Skiena's learned tome, 'The Algorithmic Design Manual' – but be warned, it's not easy reading. Algos such as 'Linear Search,' 'Bubble Sort,' 'Heap Sort,' and 'Binary Search' are in the realm of the programmer and provide the backbone for software engineering (please see Bibliography).

As promised above, in this book (you may be relieved to know) we shall be solely concerned with algorithms as they apply to stock trading strategies. In Part I we deal with the Tier 1 companies (the major players) and in Part II of this book we consider how algorithmic strategies from basic to advanced may best be used, adapted, modified, created and implemented in the trading process by the individual trader.

The earliest surviving description of what we now call an 'algorithm' is in Euclid's Elements (c. 300 BC).

It provides an efficient method for computing the greatest common divisor of two numbers (GCD) making it one of the oldest numerical formulas still in common use. Euclid's algo now bears his name.

The origin of what was to become the very first algorithmic trade can be roughly traced back to the world's first hedge fund, set up by Alfred Winslow Jones in 1949, who used a strategy of balancing long and short positions simultaneously with probably a 30:70 ratio of short to long. The first stirring of quant finance . . .

In equities trading there were enthusiasts from the advent of computer availability in the early 1960s who used their computers (often clandestinely 'borrowing' some computer time from the mainframe of their day job) to analyze price movement of stocks on a long-term basis, from weeks to months.

Peter N. Haurlan, a rocket scientist in the 1960s at the Jet Propulsion Laboratory, where he projected the trajectories of satellites, is said to be one of the first to use a computer to analyze stock data (Kirkpatrick and Dahlquist, pp. 135). Combining his technical skills he began calculating exponential moving averages in stock data and later published the 'Trade Levels Reports.'

Computers came into mainstream use for block trading in the 1970s with the definition of a block trade being $1 million in value or more than 10 000 shares in the trade. Considerable controversy accompanied this advance.

The real start of true algorithmic trading as it is now perceived can be attributed to the invention of 'pair trading,' later also to be known as statistical arbitrage, or 'statarb,' (mainly to make it sound more 'cool'), by Nunzio Tartaglia who brought together at Morgan Stanley circa 1980 a multidisciplinary team of scientists headed by Gerald Bamberger.

'Pair trading' soon became hugely profitable and almost a Wall Street cult. The original team spawned many successful individuals who pioneered the intensive use of computing power to obtain a competitive edge over their colleagues. David Shaw, James Simons and a number of others' genealogy can be traced back to those pioneers at Morgan Stanley.

The 'Black Box' was born.

As computer power increased almost miraculously according to Moore's Law (speed doubles every eighteen months, and still does today, well over a third of a century after he first promulgated the bold forecast) and computers became mainstream tools, the power of computerized algorithmic trading became irresistible. This advance was coupled with the invention of Direct Market Access for non Exchange members enabling trades to be made by individual traders via their brokerages.

Soon all major trading desks were running algos.

As Wall Street (both the Buy side mutual funds etc. with their multi-trillion dollar vaults and the aggressive Sell side brokerages) soon discovered that the huge increase in computer power needed different staffing to deliver the promised Holy Grail, they pointed their recruiting machines at the top universities such as Stanford, Harvard and MIT.

The new recruits had the vague misfortune to be labelled 'quants' no matter which discipline they originated from – physics, statistics, mathematics . . .

This intellectual invasion of the financial space soon changed the cultural landscape of the trading floor. The 'high personality' trader/brokers were slowly forced to a less dominant position. Technology became all-pervasive.

2
All About Trading Algorithms You Ever Wanted to Know . . .

Q: In layman's language what are they really?

A: Algorithms are lists of steps or instructions which start with inputs and end with a desired output or result.

Q: Do I have to know much math?

A: No, but it helps. We will provide what you need for our algos in Part II of this book.

Q: What about statistics ?

A: High school level helps. Part II of the book has a chapter which covers most of what you will need.

Q: Do I need to know Excel?

A: The book will guide you through all you need to know to use the algorithm templates which are on the CD and described in detail in Part II. Excel is a most convenient workhorse and de facto standard spreadsheet.

Q: Do I need to be generally computer savvy?

A: Not that much really – basic computer literacy and ability to handle files and mouse skill. For any real equipment function malfunctions call in an IT guy to troubleshoot the problem.

Q: Do I have to understand the detailed workings of the algorithms?

A: A qualified 'no'. Of course understanding how the machine works is an asset but you can drive a car with knowing how the engine works. If you want to design algos you will need to know where the clutch is and what it does . . .

Q: Do different algorithms work better on some stocks than on others?

A: YES, the efficiency of an algo will also vary over time.

(continued)

Q: Can an algorithm go wrong?
A: Like all software is heir to, rarely when it is well designed and tested.
Q: Do I need special computers to run algorithmic trading?
A: Depends on the level you are aiming at (a draft specification for a mini trading setup is described later in Part II).
Q: How difficult is it to learn to trade with algorithmic strategies? How long will take me to become proficient and how risky is it?
A: Part II is laid out to make the learning curve easy. A couple of reasonably concentrated weeks should provide you with the basic confidence. Give yourself two months.

The next step, so-called 'paper trading' on a simulator using 'play money', will soon tell you what level you have reached and when you feel confident enough to, so to speak, take the bull by the horns and trade real money.

All trading has an element of risk. Managing and controlling risk is part of our 'stock in trade'.

Q: How much capital do I need to trade?
A: A minimum of $25 000 in your trading account is required by the SEC current regulations to provide you with a margin account.

A margin account will allow you 4:1 trading capital intraday. (You must be cashed out at the end of the day, by 4:00 pm when the NASDQ and NYSE close.)

$25 000 is the minimum level but in our experience one should count on having at least $50 000 as the minimum account.

Never trade with money you cannot afford to lose. Putting it another way, never put money at risk which would radically alter your lifestyle if you were to lose it.

Q: Do I need to trade every day?
A: Not really, but you may find that trading is extremely addictive and you may find yourself at your computer setup from the 9:30 EST Open to the 4:00 pm Close.

Some traders prefer to trade only till midday.

Q: What other asset categories will I be able to trade using the algorithms in this book?
A: This question has a number of answers. First of all is the controversy as to whether all markets exhibit the same basic principles. (We don't think so.) Next we must look at the various asset classes: e.g. futures, options, commodities, foreign exchange in detail.

From our point of view the various asset classes are all very different from each other, but with similarities which one could explore.

This book is dedicated to the American equity market, traded on NASDAQ and the NEW YORK STOCK EXCHANGE, though we are certain that much of the machinery could be adapted to other markets.

3
Algos Defined and Explained

There are many definitions of the word ‘Algorithm.’ Here are a spray of examples:

- A plan consisting of a number of steps precisely setting out a sequence of actions to achieve a defined task. The basic algo is deterministic, giving the same results from the same inputs every time.
- A precise step-by-step plan for a computational procedure that begins with an input value and yields an output value.
- A computational procedure that takes values as input and produces values as output.

Here we should mention ‘parameters.’ These are values usually set by the trader, which the algo uses in its calculations.

In rare cases the parameters are ‘adaptive’ and are calculated by the algo itself from inputs received.

The right parameter setting is a key concept in algorithmic trading. It makes all the difference between winning or losing trades. More on this later in Part II of the book.

Unconsciously we create little algorithms without having any recognition that we are performing mathematical applications all day long. The brain supercomputer carries it all out without us being aware of it to the slightest degree.

Now let’s finally get back to trading. Here is an over-simplified algo example.

You want to buy 1000 shares of Apple (ticker symbol AAPL) and you are looking at a real-time data feed. The Time and Sale is printing mostly 100 volume lots hovering between $178.50 and $179.00 – but a few minutes ago it dipped to $177.00. So you decide to set your Buy algo the task: BUY 1000 shares AAPL at MARKET if trade price touches $177.00

Now for a slightly more complex example for which we would need a number of components. For the moment, just imagine these:

A real-time data feed (not from one of the 15 minutes’ delayed variants). This feed consists of the stock ticker symbol to identify it, the timestamp of when the trade was executed, the number of shares (volume) which has changed hands and finally the trade price as matched up by the buyer and seller who may be represented by their respective brokerages. All this happens in what we call the 'electronic pit.'

The 'electronic pit' image (thousands of traders who at that instant are looking at exactly the same data on their screens that you are also looking at) we find exceptionally useful in visualizing the price movement of a stock.

In our application a fully fledged real-time data feed is fed to an Excel template populating an Excel Spreadsheet. The template has an embedded set of Excel function language calculations (basically an algo) which Excel recomputes every time new data comes in. The algo is designed to 'trigger' when a certain calculation parameter attains a 'BUY' condition. You see this on the spreadsheet and put on the trade manually using your order management system (OMS).

In the future, we may be able to get it all done with a fully automated software subroutine with the computer taking on the order placement task for the individual trader single-handed, just as now performed by the big players of the moment!

We have purposely left the placing of orders in manual so as to accelerate the learning process and give you a firm foundation to build on.

As we delve deeper you will find that the parameter setting is, as already mentioned, one of the most crucial steps to profitability and the most difficult one to master, requiring beside experience and skill, a good helping of old-fashioned trial and error, or better yet, trial and success.

The next most important step to achieve profitable trading is to put on a protective stop loss order under every trade. This is a proprietary 'adaptive' algo which is calculated as soon as the trade has been completed. We cannot stress this enough. In an automated system it is placed within milliseconds of the actual order. With our manual system we will be a bit slower, but nevertheless it is an essential component.

The range of complexity and functionality of algorithms is only limited by the cunning of the strategists and designers. Anything they can think up can be transformed into a trading algorithm. From the most basic (e.g. If trade price of XYZ touches \$nn.nn place a market order for 1000 shares) to the most advanced which would require several pages to describe even in outline . . .

4
Who Uses and Provides Algos

As of 2009 algos have become pervasive in the financial industry.

What used to be the exclusive purview of the large Sell side firms, the Tier 1 brokerages, such as Goldman Sachs, Morgan Stanley, the big banks such as Citicorp, Credit Suisse and UBS, has now migrated even to the Buy side such as Fidelity. All are very actively pursuing algorithmic computerized trading strategies. We have selected some of the main Tier1 companies which seem to have a lead at the moment. Their lists of products are described in Appendix A.

The secretive hedge funds are one of the larger users of algorithms as these can provide substantial competitive advantage to their trading operations. As there is currently little regulation they are not required to report their activities. There is little information available regarding their operations.

The dream has always been to develop an algo which works at a high success percentage and thus is capable of providing exceptional returns. This attracts constant development investment and is vigorously secrecy-protected.

Let's take, for example, two major hedge funds such as J. Simons' Renaissance and D.E. Shaw's funds which generally produce extraordinary returns on capital year in, year out. It is rumored that each of these highly successful operations employ a staff of something over 50 PhD-level mathematicians, statisticians and physicists and run some of the most powerful and advanced computer hardware.

For this caliber of talent the complexities of an evolving market pose an insatiable challenge laced with substantial financial rewards. Here we see one brilliant exceptional individual driving the enterprise. Hardly any information is available as to their methods and strategies and actual algorithms.

The major banks and brokerages have recognized quantitative algorithmic trading as one of their major competitive advantages. These firms are all shifting financial and human resources to algorithmic trading. In Tier 1 companies the entire process is more complicated as they invariably have to deploy whole teams on a hierarchically structured basis to attain the intellectual and performance critical mass of the 'superquants' such as Shaw and Simons.

Proprietary algos and their source code are guarded like diamonds. The source code of even mainstream algos goes into 'cyber vaults' as all users have their own implementation to keep secret. Even algos in mainstream use have company-specific implementations with tweaks which make them work a bit better or do their specific job faster, or perhaps carry out some added user-specific function.

The Sell side is forced to be a bit more cooperative in disclosure to their Buy side clients. The disintermediation specter looms over the relationship (where the Buy side would take on the entire job, thus eliminating the Sell side from the game), but considering the possible financial consequences neither side is likely to give up very easily.

Whenever the markets perform their unpredictable convulsions there is always a strong move to place the blame on the latest trading strategies. The 1987 fiasco was blamed (in our humble opinion quite unjustly) on so-called 'program trading.'

The latest near-meltdown is no exception and the outcry to impose draconian regulation on hedge funds has been strident.

It is at present unclear how much regulation is going to be instituted by the SEC and other regulatory bodies and how it will operate. The very much desired disclosure of actual trading methods is highly unlikely to take place as that constitutes a hard won and highly capital-intensive competitive edge which would totally erode in value when disclosed.

The markets are in a constant state of adaptation and evolution. A prime example is the sudden widespread appearance of 'dark pools.'

Recent deregulation has allowed off-exchange trading and as the markets fragmented on both sides of the Atlantic has given rise to a rush to create Alternative Trading Facilities (ATFs) which are basically anonymous pools of liquidity.

A 'Dark Pool,' as it is so romantically termed, is an electronic marketplace that gives institutional investors the possibility to trade large numbers of shares in liquid or illiquid stocks without revealing themselves to the 'lit' market. This new area to trade has sprung up out of the deregulation changes which have been implemented both in Europe and in the USA. These 'Dark Pools' are also called 'Multilateral Trading Facilities' (MTFs) and can trade stocks listed on the 'lit' Exchanges.

Smart Order Routing (SOR) algorithms have appeared over the past 18 months and have had a fair impact on the way people trade. The SOR will choose a route primarily so that the algorithmic orders go to venues where there is liquidity, possible trading fee reduction, as well as anonymity.

The fragmentation of the markets away from the primary Exchanges (the 'lit' Exchanges like NASDAQ and NYSE) to the aggressively competing 'dark liquidity venues' is thus another area for the use of algorithms in the order routing process. These venues are in our opinion poorly policed in many cases, which allows participants to be 'gamed' quite frequently (this could take the form of 'front running' of a large order if the security is breached). The operators of these dark venues will have to take rapid and decisive action to avoid the major regulators stepping in.

There is intense activity to develop suitable algos for the optimal routing, sizing, and order process, how to split up the order and where to allocate it.

On the individual trader side of the equation there is very little data on the number, activity and distribution of activity of individual traders. One source places it at over eight million individual traders in the USA who directly trade their own accounts. There appear to be no statistics on frequency or the amount of trading these individual traders undertake. The use of algorithmic techniques is probably very small. Yet the 'trading method teaching and stock information' business for individuals appears to be going strong.

Slowly, hopefully with a small contribution from this book describing our algos and making them more accessible to the individual traders, they may be incentivized to take up the challenge and the opportunities of the advancing technologies and enter the algorithmic battle.

Algos will in any case filter through to the sole trader who will learn not only to use what there is already on offer, but will have a clearer understanding of the cogs and wheels. It is our belief this will help unleash the creativity of the great number of talented individual traders to create their own algos and encourage them to trade. It is the rate of this diffusion process which we would like to speed up and facilitate with this book and possibly others in this genre to follow shortly.

Certainly the very next phase in this development process is to bring in full computer automation for the individual trader – where the computer places the trade under full algo control. This should help to level the playing field between the Tier 1 market participant and the individual trader of modest means. We, and presumably other participants, are actively working on this – the empowerment of the individual trader. To give the individual trader, in principle, though perhaps necessarily quite scaled down, some of the same firing power as the Tier 1 giants. The first order effect should be a widening of the trading base 'pyramid,' an improvement of the basic quality of markets, perhaps less volatility, perhaps more liquidity and generally the potential for a healthier market.

As we heavily identify with the individual trader, we have to admit that we thoroughly enjoy an opportunity to do our 'bit' to contribute to the 'levelling' of the playing field, making it equal and universal for all, where previously it has always favored the big battalions.

Appendix A will give you some flavor and idea of the magnitude of algorithmic acceptance in the Tier 1 sector. It is a non-comprehensive list of some of the major players and their product offerings to the market. Hopefully the descriptions may ignite a creative spark which the reader may develop into the next 'killer algo.'

5
Why Have They Become Mainstream so Quickly?

Wall Street is usually very quick to latch on to any possible edge which would give them competitive advantage but the cross-industry adoption of algorithmic trading was even more rapid than the usual fad proliferation.

Let us consider what the reasons for this explosion might be. First technical reasons then conceptual reasons. In our opinion, having observed market behavior for nearly 35 years, this appears to be a natural evolution which typically 'catches fire' when a number of component technologies and disciplines go past the 'tipping point.'

One of the main technology components is, of course, computational power. This has behaved as per Moore's Law and has made today's standard desk top computers almost as powerful as the NASA equipment which put men on the moon and brought them back!

The telecom side has improved dramatically with much of the global system network being served by fiber optics and generally the bandwidth of transmission capability has kept pace with the exponentially growing requirements.

Various other 'infrastructure' industries have also contributed to the reliability and stability of the massive increase in the volume of trading. Algos contribute to the bottom line in a variety of ways which, when analyzed from any cost performance point of view, become a 'must have.'

However, we must not underestimate the problem of implementing algorithmic strategies. In a full bore Tier 1 company it takes strong leadership and a healthy supply of cash to put together the team of traders, technologists, quantitative market analysts and software programmers as well as the requisite IT complement. And then to mold it into an operational machine which not only can deliver the goods now but is aware enough of the evolutionary forces it operates under to be able to adapt as needed to keep ahead of the wave of competition and regulatory changes.

The vogue for team-oriented design is especially hard to implement in the political climate of a large corporation.

Here is a mini mythical scenario.

A designer comes up with a good idea for a complex strategy. This may need lots of thought and can be mulled over and over, sometimes for weeks or even months by one or more designers. Rarely, it can be a team effort with representatives from the various departments taking a hand. Often the CEO has a strong hand in the vision.

When the designers are satisfied they have something of interest they will turn the new algo specification over to the programming team so it can then be converted into executable software, coded and optimized for speed and reliability.

Then comes another level of the dreaded 'back test' to see how the new algo performs on historical data. The efficacy of the algorithm over a period of weeks or even months (in some cases years) of historical data is compared against various benchmarks. (See Part II for standard and proprietary benchmarks.)

More thorough testing takes place before it goes 'live,' to ensure that the algo absolutely does what it is intended to do, and nothing else, under all circumstances and market conditions which it can possibly encounter.

It can then finally be deployed on the trading desk and carry out the many man-months of work in a flash, over even thousands of stocks – far beyond the reaction time, concentration ability, and reach of a human trader. So the prethought and tested concept is amplified multifold and 'goes into production.'

Of course the CEO and Head Trader and the whole team will be involved in further improvements and refinements which come to light as reality hits the algorithmic design.

However, the benefits are immediate:

- Cost per trade reduction is substantial.
- General throughput speed is increased and thus more business can be transacted.
- Self-documenting trade trail meets financial control and regulatory requirements.
- Reduction in trading errors.
- Consistency of performance.
- Less trading staff 'burnout' as the emotional side of trading is dramatically reduced.

The strategic reasons are also fairly obvious. The exponential increase in trading load made the 'just add more traders and more trading floors' option unattractive, and perhaps even unsustainable from a cost, staffing, security, performance, control and efficiency point of view.

Here computers come to the fore as there is virtually no limit to their capacity and patience. You can set up a watchloop on the streaming trade data of a ticker symbol and the machine will obediently watch for the parameter you have set – and watch, and watch . . . without getting tired, bored, losing concentration, forgetting what it is supposed to do . . . and then, finally when (and if) the parameter is hit . . . Wham! It executes your instructions.

Enter the Algo Age!

6
Currently Popular Algos

The following is a very sparse list of mainstream algos currently in common use (mid-2010). One should point out that the prime function of most current algos is to 'get' the trade without market impact, anonymously, rapidly, without being 'front run.' Making an immediate profit on the trade is not of first importance to Tier 1 companies as most of these trades are of longer duration. Only the so-called 'high frequency' traders who use the technology to their main advantage by minimizing the ping time to the Exchanges and are happy with a couple of basis points per trade are 'immediate profit oriented.' Their operations are held strictly confidential and have caused a certain amount of controversy and regulatory interest.

Algos for the use of individual traders are designed to provide immediate returns. Part II of this book describes some of these in detail.

VWAP – Volume Weighted Average Price

This is probably the oldest and most used algo. It is often used as a benchmark by the Buy side placing orders with the Sell side. We shall therefore provide a more detailed treatment of VWAP than of the other algos that follow. It has a large number of variations designed to accomplish specific tasks and we shall concentrate on it to explain some of the various tweaks and variants that may be implemented.

The VWAP engine uses real-time and historic volume data as a criterion to size the slicing up of large orders over a set period of time or throughout the trading session with respect to the stock's liquidity. The trader specifies the number of discrete time intervals (sometime called waves) for the algo to trade a quantity of shares which is directly proportional to the market volume in the time slice.

Often the main challenge is to make trades throughout the day which track the VWAP. With such orders, which must be worked over several hours, the automation of a VWAP algo provides meaningful manpower cost savings. Therefore the VWAP strategy is most often used on longer duration orders.

VWAP is very often used as a benchmark for block trades between the Buy side and the Sell side.

The slicing up of large orders into many smaller ones improves the chances of reducing the risk of market impact cost. It also helps making the order size invisible to other market participants.

The volume profile calculation and prediction together with the real-time actual volume govern the size and frequency of the orders put on by this strategy. The frequency is worked so as not to be 'recognizable' by a market competitor, as is the volume of each order wave to reduce the chance of being 'front run.'

Due to stylistic and ergonomic features of the market it is not unusual for more volume to be traded during the early part and latter part of the session in order not to create an adverse impact on the price.

The basic formula to calculate VWAP is:

$$P_{vwap} = \Sigma(P * V)/\Sigma V$$

where

P_{vwap} = volume weighted average price
P = Price of each trade and
V = is the volume of each trade.

This basic algo has many variations which have been developed by the Tier 1 users over time, some tweaks being proprietary to specific entities and held very confidential. For example what lookback period we use for the VWAP calculation and various 'tweaks' to depart from the plain vanilla algo such as constraining execution price, or, more frequently constraining volume.

VWAP strategies may be implemented in a number of ways. The order may be sold to a broker who will guarantee VWAP execution on the day. (He will charge a fixed agreed upon commission in return.) Another way would be to take the trade directly through an automated participation trading algo which will slice up the trade and participate proportionately to the current volume in the market with hopefully as little market impact cost as possible. Again one must be careful of being 'discovered' and front run. More variations are constantly being tested and the basic implementation is being refined.

Orders can be sent to the market according to a preselected strategy – for example we can send waves into the market according to the well-known daily 'volume smile' where there is more activity at the start and at the end of the trading session.

In all cases we must be aware that we are dealing with a moving target – the volume pattern of a stock on any particular day may vary substantially from its average. It even depends on what type of average we use, and how long its lookback period is.

The volume distribution time function varies considerably between different stocks – more variation is experienced with thinly traded stocks both intraday (during the course of the trading session) and EOD (End Of Day), and predicting anything from historical volume data for a thinly traded stock is a dicey enterprise indeed.

For those interested in the source code for VWAP, there is an article by Andrew Peters on the website www.codeproject.com which gives an interpretation. As we have mentioned, the actual code on any algo implementation used by any particular Tier 1 company is a well-kept secret as each company invariably makes small changes to suit their individual requirements as well as to improve the anonymity of the algo.

TWAP – Time Weighted Average Price

This algo strategy simply divides the order more or less evenly over a user specified time frame. Usually the order is sliced up equally into a specified number of discrete time intervals, or waves. Though convenient this may expose the trader to other market participants' 'sniffer' algos which search for just this kind of predictability in a competitor's trading and quickly take advantage of it by 'front running' it. This is often combated by leaving out a wave or using a 'fuzzy' time interval spacing or even a 'fuzzy' number of shares per wave. The more sophisticated (or more paranoid) trading desks use a random number generator . . .

POV – Percentage of Volume

The main target here is to 'stay under the radar' while participating in the volume at a low enough percentage of the current volume not to be 'seen' by the rest of the market. The rate of execution to trade up to the order quantity total is kept proportional to the volume that is actually trading in the market. This provides a certain amount of 'cover' especially when trading a large quantity of shares.

'Black Lance' – Search for Liquidity

This menacingly named algo is designed to find liquidity in so-called 'Dark Pools.' This is accomplished by 'pinging' the many different venues and analyzing the responses to determine the level of liquidity available in the issue of interest.

The Peg – Stay Parallel with the Market

The PEG algo sends out limit orders, randomizing the fraction of the total order and follows the market, very similarly to a trailing order.

Iceberg – Large Order Hiding

Here we try to hide a large order from the other market participants to avoid them 'front running' it and generally to minimize market impact cost when we are trying to accumulate a significant position in a particular stock. This is done by slicing the order into many smaller segments and randomizing the order placement. Smaller orders hopefully improve the chance of avoiding market impact cost. There is a limit order version of Iceberg which can be deployed over longer time periods.

Most of these major algos are used by institutions shifting huge amounts of stock. Just making the trade safely without being 'front run' takes priority over making immediate profitable returns. Anonymity rules.

This is in total contrast to the individual trader with limited capital where immediate profit has to be realized and the size of trades is small, in the region of 1000 shares per trade with a very occasional 2500.

Algos for the individual trader are therefore quite different from those we have described above. We call them ALPHA ALGOS and, as already mentioned, we shall describe exactly what we mean by that in Part II where we shall be dealing exclusively with these algos which are specifically designed for the individual trader.

There are a large number of variations derived from the basic trading algos. Some trading problems are more amenable for solution by one algo variant than another, some can only be solved by a very specific algo implementation. We have selected from the multitude a few algos just for reference and to feed the creative instincts. Here are a few more.

Recursive Algos

Recursive algos 'call' themselves over and over again until a preset condition is met. We are told that the famous Buddhist 'Towers of Hanoi' can be solved by a recursive algo. We have not tried it ... as myth has it that when the monk in charge of the golden disks in a Buddhist monastery moves the last disk over to its final resting place this will signify the end of the world. Can't risk it!

Serial Algos

These algos are really a set series which executes on a computer serially, the algos are set up to follow one another possibly with some logic control to provide branching.

Parallel Algos

These algos make use of multi-core computer architecture and are able to execute multiple instructions simultaneously.

Iterative Algos

These use repetitive constructs like 'if ... then,' 'Do while,' 'for ... Next' to control execution flow similar to those available in programming languages, usually with some parameterizable values to test and make 'decisions.'

Pair Trading Strategy

We will now have a look, in some detail, at pair trading strategies which initially provided the main impetus to algorithmic trading. The fundamental one we shall describe bears little resemblance to the sophisticated versions used by statistical arbitrageurs.

As volatility has increased not only at the individual stock level but also in the overall market swings becoming more drastic in the recent past, even without the earthquakes of potential market meltdowns which we have experienced in 2008, it has again become very attractive to find a 'market neutral' strategy.

Pair trading is market-neutral by its very structure as its operation does not depend on the direction of the market but on the correlative and anticorrelative behavior of the stock being traded.

Nunzio Tartaglia, while at Morgan Stanley, found that certain stocks, usually in the same sector and industry, showed strong correlation in their price movement. He therefore reasoned that any departure in the pattern of co-movement of the two stock prices would revert back to 'normal' over time. This conjecture proved to be correct and was worth a king's ransom to Wall Street for the next 20 years.

When the correlated stocks started moving in opposite directions the strategy was simply to short the out-performing one and long the under-performing one.

The conjecture continues with the thought that the two stocks will exhibit a reversion to the mean type behavior and converge back to moving parallel. When this happens we obviously close out the trade with two profits.

An important point often missed in this strategy is that it is rock solid market-neutral as we have a short and a long position in place so that whatever the market does we are immune to it. (One of the occasions where risk again enters the picture with a vengeance is when one of the pair suffers a liquidity crisis and you cannot close out the trade.)

To trade a pair strategy we have to find two stocks which are highly correlated over our lookback period. The algo will watch out for their movement out of the security envelope and once the deviation crosses the trader's preset limit will go into action. It will monitor progress (sometimes for quite an extended time period, such as days or even weeks) and when the prices revert to their normal correlation it will close out the trade.

Let us always remember to give credit where it is due: The pair trading strategy, as previously mentioned, was designed by a team of scientists from different focus areas such as mathematics, computer sciences, physics, etc. who were brought together by the Wall Street quant Nunzio Tartaglia who was the main driving force of the team.

In the 1980s Gerald Bamberger popularized this strategy as he headed the team of quants at Morgan Stanley. His team, along with Nunzio Tartaglia, proved beyond any reasonable doubt that certain securities, often competitors in the same sector, were correlated in their day-to-day price movement and they started putting their money on it with incredible success.

A useful website which has correlation information for use in the selection of pair trading candidates and also shows the topological relationship between stocks researched by Professor Vandewalle at the University of Liege is: www.impactopia.com. We have found that Vandewalle's topological displays of stocks provide a visual map similar in many cases to our cluster analyses. XY scatter charts add clarity. This can be helpful in selecting cohorts of similar stocks with the conjecture that stocks with similar metrics will trade in a similar fashion.

7
A Perspective View From a Tier 1 Company

UBS has been one of the early major players since circa 1990. The then chairman of UBS, Marcel Ospel, took a great personal interest in the work involving chaos theory and financial prediction of Professors Farmer and Packard and their company located in Santa Fe, New Mexico – The Prediction Company. (In the book of the same name John Bass chronicles the startup of the company – please see our Bibliography.)

This interest resulted in an exclusive contract for financial work with UBS and over the course of time The Prediction Company was eventually bought by UBS outright.

Here we would like to quote an article by Mr Owain Self, Executive Director, European Algorithmic Trading at UBS, which we found most enlightening and which we feel puts the case of algorithmic trading extremely clearly, and comprehensively. Our thanks go to Mr Self and UBS.

BUILDING AND MAINTAINING AN ALGORITHMIC TRADING TOOL

Owain Self
Executive Director, European Algorithmic Trading, UBS

As the use of algorithmic trading tools has become more widespread, clients' expectations have similarly increased.

Owain Self of UBS highlights the important considerations when developing and supporting a credible algorithmic trading tool and what it takes to maintain that credibility in a constantly changing market.

In the fledgling days of algorithmic trading it was possible to buy or to build a system on a very affordable basis.

But, as usually happens, it was not long before the realisation hits that you would get what you had paid for.

Clients' expectations have since changed. The initial frenzy among broker-dealers to have any kind of algorithmic capability within their portfolio of trading tools, regardless of the robustness or performance of the system, has given way to an increased level of circumspection. And those providers that were offering affordable but less effective systems have been found out.

In today's markets, clients are looking for performance, flexibility and reliability – attributes which require an investment running into the tens of millions and a worldwide team of people that exceeds 100. This realisation has limited the market to a select and credible group of five to six major, global broker-dealers who are willing to make this investment in technology and expertise. But what does it take to reach that elite group of providers, how should that investment be spent and what work is needed, in a trading discipline where performance and capability must constantly be improving, to maintain a position at the top of the algorithmic trading table?

Expertise

The first investment must be in assembling a team of highly qualified experts. There are three areas to draw from – the traders, the quantitative analysts and the technology developers – and it is essential to create a balance between these three groups. It is unrealistic to expect to be able to recruit an all-rounder who is the best-in-class for all three groups and it is also unwise to invest too strongly in bringing in a technology capability without having a similar pool of talent in the trading and quantitative areas.

There are clear responsibilities within these three groups – the traders will be more experienced in the end-users' behaviour and what they would like to see in terms of functionality. However, this does not mean that the development process should be reduced to the traders producing a list of functions and features and then expecting delivery of their dream some days later.

The quantitative analysts are becoming increasingly more important in the development of algorithms, as the models and the understanding of risk takes on more sophistication but a successful system will not be one that runs the most mathematically sophisticated modelling process but is nigh on impossible to be used by the average trader.

Everybody has to be involved in the decision-making process.

At UBS, and as a global investment bank, we are able to draw from our considerable skill base in-house and select people that show the right blend of expertise that can be brought into a team that will develop a process where input comes from all three sides – the technology, the traders and the quantitative analysts.

Technology

Development of the original system is only the beginning of the project and at UBS it is difficult to pinpoint that singular moment of genesis. It has been continual progression from the late 1990s onwards to the point where we are now on our third generation of algorithms.

Technology obviously plays an essential part in a process so dominated by performance and a considerable investment has to be made. The latest generation of technology is essential and legacy systems are to be avoided – particularly when constant development and improvement plays such an integral role in the product's success.

Whereas the original algorithms in those first generation products in the late 1990s were more rudimentary and could theoretically have been developed by a trader, the ongoing sophistication in trading and modelling means that there is now far more input from quantitative analysts and a necessity to find the most suitable technology as opposed to a standard software package or an off-the-shelf piece of hardware. Each new generation of algorithm trading tools, however, is not built from scratch. It is about finding the limitations of existing systems and then making the necessary improvements so it is often a case of moving sideways in order to go forwards.

Feedback

Changes in the marketplace also necessitate constant redevelopment and for this client feedback plays a vital role in development efforts. However, to ensure that the maximum benefit is derived from the feedback, it is important to look beyond simply acting on verbatim customer comments without setting them in any kind of context. The level of expectation and of education must be considered alongside any clients' comments. For example, a client may not appreciate the way a process is performed because it is simply not the way they are used to working – which does not necessarily mean that the process is wrong or in any way inferior. Working constantly with the client so that there is a true partnership rather than a one dimensional client/vendor relationship helps to determine what feedback can be useful for future development as opposed to a 'what the client wants the client gets' dynamic.

It is also important to rely not solely on your customers for their feedback. By sitting waiting for this feedback to come in, the lead time involved would mean that none of these changes would be implemented on time. We would also all be developing the same product because every provider would know what these clients' demands were. In a fast moving environment such as algorithmic trading it is vital to stay one step ahead not only of competing providers but also in trying to anticipate your own clients' expectations and feedback. This can be generated internally from the banks' own internal trading teams and from the development team itself.

Other considerations

While the size of investment and the level of expertise are strong determining factors in the success of an algorithmic trading product, there are other factors to consider. The ability to fulfil all of a client's trading ambitions is as important as is the level of customer service in that there is someone on the end of a phone for clients at all times. An algorithmic trading tool is not a stand-alone system so it has to be able to integrate with other systems within the trading departments and also the back and middle-office processes. Global reach is also important for those operating in multiple jurisdictions. During the development phase we like to work with regional experts when looking at products for individual markets because they understand those markets better – as opposed to developing one monolithic product that you then try to adjust for each market. It also important to understand and appreciate the regulatory developments that are taking place in each region; that it is possible to build any of these changes, as well as general changes in trading strategy, into an algorithmic tool. For example, with MiFID taking effect in Europe and similar developments happening in parts of the US market, such as Reg NMS, traders will be looking to access multiple venues. Speed is important to a degree, although not as important as say direct market access in terms of execution.

But the fact that UBS is an exchange member in most markets is a big advantage over some brokers' offering algorithmic tools. The final advantage is the anonymity which clients have using the trading system and the reduction of information leakage. It is important that the technology and controls are there to ensure this level of security while still maximising crossing opportunities within the investment bank. Clients' confidence in their broker increases when that broker is more open how internal trading desks use the algorithmic trading systems. Whether the user is an internal or external client, their access should be the same.

It is a constantly evolving process and once an algorithmic trading system has been built, one can never sit back and rest on contented laurels. The development is constant. It is day by day, minute by minute and millisecond by millisecond. That vested effort involved in tracking market and regulatory development and clients' requirements will never slow down.'

8
How to Use Algos for Individual Traders

So how do we, the individual traders, go about using algorithmic trading technology to improve our trading results?

ALPHA ALGO

The generic term we coined for most of our work is that of 'ALPHA ALGO' to flag that we are going primarily after profit per trade.

Liquidity

We, as individual traders are basically not concerned with liquidity as the stocks we have selected to trade are virtually always liquid well over our trading threshold. Remember that we trade small amounts of stock – as mentioned above – say 1000 to 2500 shares at a time. This is something like one or two orders of magnitude smaller than some of the trades put on by Tier 1 institutions (at least before their 'slicing machinery' gets to work). They would see us only as 'noise' in the system.

Avoid

Stocks which are thinly traded, say less than 500 000 shares per session.

Manual trading

The example algorithms described in Part II of this book are designed to be used by the individual trader and traded manually. We believe this route to be the best way to gain a bedrock understanding and feel for how the algo-market-stock triangle interacts.

Once you have worked through the Excel templates and the explanations you should have a workable idea of what there is that you can implement and how to go about it.

Simulators

Practice on a trade simulator (e.g. TALX from TerraNova Financial Group, or the Ameritrade simulator) is a good lead-in to actually using live bullets. You should be able to run the SIM without hesitation or having to think about it.

Parameterization

As is the case with all algorithms ALPHA ALGOS need meticulous parameterization (setting of various constant values in the algos' formulas) in order to achieve good results. The lookback period may vary over time for any one given stock and needs to be adjusted periodically, driven by the changes in activity of the stock. Ideally parameterization should be performed, preferably daily, using a lookback period matching the profile of the stock. Shorter in our view is better. We find a five-session lookback plenty in most cases.

Ensure you are Comfortable with your Understanding of the Algo you are Using

A comprehensive understanding of all the components which go into the construction of the algo, its parameters, strengths and limitations will give you the basic confidence and reduce, if not completely remove, the emotional burden in trading.

So putting on a trade becomes less emotional, as raw skill, market 'feel,' experience, confidence and daring are to a large extent replaced by your computer and the 'prethought' software running the algo. We cannot emphasize enough that you are using prepackaged thinking time.

Do not Trade if you do not Feel Well, Physically or Mentally

Be that as it may, we still do not recommend that you let the algos loose when you do not feel in top form. Like Scarlett said in *Gone with the Wind*: 'Tomorrow is another day'

Do not 'Second Guess' your Algos

Once you have selected your 'Watchlist' of stocks to trade and the matching algorithms let the computation take the strain and do not even be tempted to second guess them. As already mentioned, this removes the draining emotion from trading and gives you a much clearer view and understanding.

Algos are Power Tools – Mastery Takes Practice

Algos are power tools and results obtained are directly proportional to the power of the tools as well as the input of abilities, ambitions and goals of the user with the market evolution acting as the 'random' variable.

As with all skills it takes practice to attain the requisite level of mastery over the material and equipment so patience, perseverance and lots of practice will bring results. You should budget enough time to trial various platforms to select the one that suits you best.

After a certain amount of time most of the interactions between the individual trader and the OMS (Order Management System) become automatic or semi-automatic as is the usual course in learning any intellectual/motor skill. The time this takes varies greatly between individuals – from several weeks to several months is not unusual.

9
How to Optimize Individual Trader Algos

We have found that much of the overall optimization of individual trading results lies in the choice of the stocks to be traded.

The second in importance is the judicious setting of the required parameters, followed closely by the ALGO/STOCK tuple match.

Here are a few rough guidelines we use for choosing our stocks. These are designed for the individual trader and are strictly our own preferences. You will soon develop some favorites and our bet is that these will provide you with the bulk of your profits.

1. Volume > (greater) than 1 million shares over session.
2. Trade price > (greater) than \$35. We generally prefer the higher priced stocks. LC volatility index > n (please see 'METRICS' file on the CD).
3. Parameter optimization can be carried out by backtesting; we use five consecutive sessions as our lookback default in most cases. For very high volume stocks (>15m) increase to 10 trading sessions.

Dealing with the Tier 1 algos the optimization parameters are of a different nature just as the goals of these algos are different from those of the ALPHA ALGOS for the individual trader.

For the Tier 1 institution much is made of execution speed, dark liquidity access, order routing, disguise, anonymity, secrecy and completion of the order. We shall leave these subjects to the Tier 1 designers.

For our individual trader ALPHA ALGOS: We need ALPHA, we need positive cashflow. If the standard benchmark is positive we may possibly take something of a free ride because as they say: 'A rising tide lifts all the boats in the harbour.'

Once we have something that works on backtest consistently (say six out of ten) with a bit of optimization by varying the key parameters plus/minus 5% we may have reached say seven out of ten trades profitable, which is quite a lot better than what most top traders are able to routinely achieve. We can, if we wish do some more optimization. It is hard to guess where we reach the point of diminishing return for our efforts.

We can optimize on 'basis points per trade second' as the main optimization criterion. In other words per cent profit per second, which includes a bit of guidance as to risk. $1 on $100 share in 30 seconds is better than the same in 60 seconds. $1 on $50 share in 30 seconds is better than both the above.

In principle, the longer your money is exposed to the market the longer the time that it is at risk, not in your control. The longer the money is tied up in a trade the longer is the period when you cannot use it on another trade.

So, ideally, if your trade is producing a predefined return as measured by total basis points per second for the trade, you hold the position. As soon as the basis points return wanes past a set threshold you put in your sell at market order (do not forget to cancel your protective stop first).

You will soon be able to make these mini calculations unconsciously. It will become second nature.

The much maligned word 'Churn' is now the key optimizing parameter as we need the capital to produce the requisite returns. So if a stock or algo routinely underperforms it is put on the back burner for analysis on how its algos can be brought up to par.

Optimization here is a process of trial and error and certainly not in any way 'rigorous.' It is a moving target which you try to reach. Let's retrace back to the optimization effort.

We usually start to test an algo frequently (every few sessions) by varying the lookback period of the moving averages to see if any small increase or decrease makes much of a difference. Then we vary the SMA and LMA or EMA parameters. It's quite important to keep a log of what has been varied and what the results were and when.

The next step is a look at the trigger parameters, for example starting with the valley negative values and the ALPHA constant in EMAs. The ALPHA constant in EMAs controls how far back the algo looks for influence from the price tick series. Vary ALPHA by 100T increments to observe the effect this has on the triggers produced and the ensuing trades.

This whole process is at first quite tedious but gets easier with a bit of practice, and some good results to keep the enthusiasm up.

An optimal ALPHA ALGO target is eight profitable trades out of ten, with two stopped out by our proprietary stop loss algo (which is designed to cut in at about 40 basis points, with some adaptation driven by a lookback period calculation).

The profitable trades should average between 25 and 40 basis points net of commission.

Do not be too worried if you do not reach these targets straight away as it will take a while to get comfortable with all the items that go into putting on trades – the

Order Management System (OMS), the speed with which you can comprehend the real-time charts and the confidence in putting on trades without hesitation; even more important, how you parameterize the individual stocks you trade and obviously, most important of all, your choice of stocks to trade.

- Be slow and methodical.
- Keep as many or as few paper records as you are comfortable with, the rest can stay on your OMS.
- You should download the session records to a log file diary for future reference.
- A backup of the file once a week is not a bad idea.
- Keep a pad and pencil at hand all the time to jot down ideas which often come in the heat of battle.
- Keeping a pad and pencil by your bed is useful for those ideas you get in the middle of the night and promise to remember.
- Do not trade when you do not feel well or emotionally upset or have not had enough sleep. A missed keystroke costs . . .
- Do not try to hit home runs – our whole philosophy works on plenty of fast singles and doubles.

Wall Street sometimes has a useful maxim: 'There are Bears, and there are Bulls. Pigs get slaughtered.'

10 The Future – Where Do We Go from Here?

The future is most decidedly pan-algorithmic. The Sell side brokerages are virtually turning into software houses (sorry for slight exaggeration) being in competition with each other for the Buy side billions (or is it already trillions?).

In turn, the Buy side is experimenting with doing their own thing and developing their own proprietary algos and trading desks to run them. This is a potential scenario for disintermediation (the skipping of one or more members of the marketing chain). The Sell side brokers are aware of the problem and aggressively develop algos to the customers' orders to keep the relationship going in the right direction and maintain the status quo.

The human trader's attention span, processing and endurance just cannot match the performance requirements of the explosive developments in trade volume and market activity and complexity which is increasing exponentially.

The Tier 1 companies will increasingly have to rely on algorithmic trading and will shrink their trader headcount in the process, as the avalanche of information to be processed and reacted to goes from gigabytes to terabytes to petabytes.

The massive investment in algorithmic trading will provide many surprises – 'intelligent' algorithms which make their 'own' decisions to optimize trades.

The area of artificial intelligence has not been sufficiently explored and will eventually yield some major advances in the near future. Predictive algorithms will emerge with a reasonable hit rate and the predictive horizon will stretch further into the future.

As Artificial Intelligence (AI) and Artificial Neural Networks and Genetic Networks develop to reach critical mass with the increased capability of predictive estimators we envisage a shoal of rapacious ALPHA ALGOS will be released into the algorithmic trading space. In our miniscule way we are the vanguard providing the initial dose of ALPHA ALGOS for the individual trader. We will be researching some AI and Artificial Neural Networks to extend the trading arsenal of the individual trader.

We can imagine computers with a Master Algo deploying dozens of specialized algos matched to the personality profiles of individual stocks with real-time backtesting and parameter setting – perhaps advising on stock selection to the Head Trader, or even taking the initiative and pricing the stocks and doing the trades on its own. Or even advising on the selection of the Head Trader.

Futurology, a pseudo science predicting the future, would add a few science fiction twists with Neural Networks and AI used not only to put on trades but to make decisions on which human traders should or should not be in charge, judging from genetic profiles, and a variety of specialized tests constantly carried out by the dreaded 'Master Trader Selection Algo.'

In the long run it looks like it will end up as a battle between supercomputers, with the individual trader almost totally out of the actual trading picture – but that may still be a while yet. There may be enough time to reverse or flatten out the trend a bit. Ray Kurzweil's vision of the 'singularity' does, however, loom on the horizon.

Such a supercomputer scenario is fraught with potential disasters of all types – regulatory, operational and human. It would be frighteningly easy to subvert such a powerful system as the architectural complexity would be beyond the intellectual grasp of any single individual, no matter how talented. The system could then succumb to a disciplined team approach.

We hope that the individual traders will rapidly engage with algorithmic methods and that their number will grow exponentially, to the point of again being a major factor in the markets and providing a counterweight to balance the powerhouse structures with the promise that some sanity and stability will be maintained.

It would be a great achievement for the markets if more and more individuals would enter the fray and take accountability for trading their own hard earned funds. To them it is their lifeblood. To the professional traders it's a well paid job.

We would like to contribute to this hoped-for trend by providing the requisite concepts, software and background information. Leveling the playing field and keeping it from tipping over is a massive task, but in our opinion, a most worthwhile one.

Part II
THE LESHIK-CRALLE TRADING METHODS

Algorithms For Individual Traders

Preface to Part II

Our trading methods took 12 years of live trading on the NASDAQ and New York Stock Exchanges and research and development in the microeconomics of markets to reach the current level. This work was carried out in almost paranoid secrecy as any 'leakage' would tend to compromise our efforts.

We worked in London, Chicago, San Francisco, Louisville, and Grand Cayman with no outsider ever seeing our computer setups or source code. Further research work on full automation of the trading methods, including multi jurisdiction trading and capital use optimization is in progress.

The methods we used from the start were strategies mathematizing our conjectures (mainly based on empirical observation and live trading) into algorithms implemented in Excel. Many of our key principles were developed to completion only in the last 12 months and are disclosed in this book in simple to understand form, leaving the more advanced rigorous math for a future volume.

Our first innovation is the almost exclusive use of 'tick series' rather than 'time series' to obtain the highest possible resolution. (More about these shortly.) The Exchanges report each trade (or tick) in real time as the trade occurs. We pick up the data from a data feed and our software writes it to our Excel templates. This asynchronous data feed is automatically more current than any aggregated feed.

It took us quite a while to get used to the 'fire hose' of data coming at us. It took Edward quite a lot longer than Jane who picks these things up on the run. It reminds her of her first ticker tape at Fahnstock and Company, her first job in 1978 which started her career. She grew up in the industry from a 17-year-old to a seasoned licensed investment professional.

Edward has been trading NYSE stocks before NASDAQ was created, carrying on a family trading tradition, both parents traded American equities after World War II.

The twelve monitor array and 18 core server we use for our research sometimes seems totally insufficient and just not up to handling and displaying as much data as we wish. One does suspect that there is also a built-in limit to how much our perceptual systems are able to take in and make sense of.

At one point we had a desperate call from the datacenter at the brokerage asking us to please reduce the number of five day tick charts we had displayed on our system (350 IS quite a lot, one must admit, but there is a saying 'if you cannot stand the heat, stay out of the kitchen') as their system was about to crash. (Yes, we took it down to a puny thirty charts.)

Due to this ultra high data resolution we can identify and take advantage of small, fast, intraday swings in price. It is a bit like looking at a specimen under a high powered microscope.

These small swings become invisible as soon as you use any data aggregation, at even one second aggregation. This phenomenon gives us a substantial playing field.

Our real-time recalculation and response time is short enough for us to be able to use market orders most of the time. It is a fair bet that most stock prices are unlikely to move very far in less than 150 milliseconds, our average recalculation time.

There are about 8000 issues quoted on the NASDAQ and NYSE Exchanges. This imposing figure however includes quite a large number of issues which trade very thinly or even not at all on any particular day.

We trade a very small selection of stocks which meet certain strict criteria on our various metrics including, among many others the very obvious volatility, volume, and price levels.

The set of our proprietary algorithms described in this book was selected for ease of use by new traders with only limited experience. The algos, templates and supporting material are provided on the accompanying CD.

We have found, anecdotally, that certain types of algo appear to provide better results with certain types of stocks and certain stock sectors. We have not been able to find any predictors to be able to reliably forecast these patterns up to this point.

The prevailing market regime and sentiment swings and regulatory changes influence the behavior of the stock-algorithm couple, as do various other stylistic properties of the markets, possibly swamping the specificity effects on which we rely to provide us with the trading information. In all likelihood, the variables on which we base our trades are of a smaller order of magnitude than many other events which are 'unobservable.'

As mentioned previously we believe that a reasonably rigorous mathematical treatment of market behavior and efficient prediction of price trajectories over a useful time horizon is still quite a number of years away.

At present, we have to be satisfied with relying on a hybrid approach: A melange of chaos theory, complexity theory, physics, mathematics, behavioral Prospect theory, empirical observation, with a great amount of trial and error experimentation, experience and creative insight thrown into the mix. This, so far has been the only way for us to achieve reasonable results and to get some rewards for our efforts.

We often use the weather analogy to explain our core philosophy: Can you tell us exactly what the weather will be three days from now? Can you tell us exactly what the weather is going to be in the next three minutes?

This naturally leads us to the fairly obvious conclusion that the prediction horizon gets fuzzy rapidly. So going to the limits of shortest time of trade and highest data resolution appears to be a reasonable strategy.

As mentioned above, in order to use the highest resolution of real-time data possible – we use the datastream reporting individual trades or 'ticks' from the NASDAQ and The New York Stock Exchange for our input data. Each trade is one line in our Excel template and reports date, time of trade, volume (how many shares traded) and the price at which they changed hands.

A 'tick' series is not a true 'time series' (the former is asynchronous, dictated by the occurrence of trades while the latter forcibly aggregates the trades into a time interval, be it minutes, days etc.).

Nevertheless we have found that much of the time series analysis machinery and other statistical weaponry can be used with 'tick series' with only minor adaptation and a modicum of hand waving.

We track the stock price movement of each individual stock we wish to trade using its own Excel template, one 'instance' of Excel for each stock. We can trade as many instances of Excel simultaneously within the limitations of our hardware, operating system and connectivity. Our maximum was reached on a fairly slow trading session where we were running 40 stocks on an 18 core server. For the starting algo trader we recommend starting with one or two stocks and a reasonable machine (see our chapter on Hardware) and then, when that feels ok add more stocks and switch to more powerful hardware as you gain confidence.

The incoming data initiates recalculation of its target spreadsheet and our algorithms are built to recognize trade triggers, usually peaks and troughs ('valleys' or 'negative peaks' if you like) within three to seven ticks of the turn. This is fast enough to put on a trade with a very high probability that the trade price will not have moved appreciably from the trigger point.

We should mention that a very quiet market can suddenly erupt like a winter squall on a frozen lake sweeping all in front of it. It appears sometimes that a long 'lull' precedes this sudden wealth of opportunity.

For a long time the conventional Wall Street 'wisdom' was that this 'fire hose' of information was mainly random noise. This view is slowly changing and is supported by the views of various highly recognized academics such as for example Dr Benoit Mandelbrot who has provided much new financial analysis.

We do not use preconceived models of stock price movement but prefer to just let the data 'speak' and direct us what to do.

The market has proved to be too multidimensional and constantly evolving as well as being subject to fairly abrupt 'regime' changes to fully surrender its secrets to any modelling technique or distribution at the current level of our understanding.

Our core strategy: Many small trades of short duration and quite small returns on each trade.

Our software uses various selected algorithms to find these small and short lived market inefficiencies where we can achieve a profitable trade. The buy, or 'long' side is the easier algo to perfect. The sell algos are much harder to get anywhere near optimum. We have on occasion had to simply use a $ or % value to trigger the sale (or the short cover) of the position.

If one thinks about it for a while the puzzle dissolves: When buying you have a huge selection to buy from. When selling you must sell just what you bought. The opportunity size difference is many orders of magnitude.

In all cases we protect the trade with a proprietary adaptive stop. This so-called 'capital protection stop' took an inordinately long time and hefty effort to develop as we originally completely misjudged and underestimated the magnitude of the problem. Only very recently (mid 2009) have we finally been satisfied with the stop loss algo and have added it to all trades.

At this high resolution (tick) the price swings are very frequent which provide us with the extremely short holding times – trades with very low duration. This frees up the capital which had been committed to the trade to be deployed on another trade. This produces what we call 'virtual margin,' providing us with an advantage to use our small capital as efficiently as possible.

In order to track, quantify and to be able to quantitatively compare these small swings we devised a 'goodness of trade' metric: 'basis points per second.' This measures the percentage profit (loss) per second on a completed trade. It can also be used as volatility metric between two designated tick points.

$$bp/sec = [((T_n - T_0)/T_0)/(d_n - d_0)] * 10\,000$$

where

T_n is the end price of the trade,
T_0 is the start price of the trade,
d_n is the end of trade time stamp
d_0 is the start of trade time stamp
$(d_n - d_0)$ = the duration of the trade in seconds
'n' is the number of ticks from start of the trade, which is designated '0'.

[A basis point is a percentage multiplied by 10 000.]
E.g. 0.01, or 1% * 10 000 = 100, thus 25 basis points = ¼% and so on

Why does our methodology facilitate a high success rate? Here is our explanation:

A so-called 'pattern trader' gets 4:1 margin on the cash in his/her brokerage trading account. The $25 000 minimum means you can trade $100 000 intraday as long as you are 'flat' (out and in cash) by the close of the session. Add to this the fact that the trades are all of quite short duration which allows the $100 000 to be traded several times during the day. We thus obtain substantial intraday leverage, which we call 'phantom leverage,' as it all vanishes at 4:00 EST as we go back into cash.

One could thus trade a multiple of $100 000 in one session. In our experience, trading churn of more than 4 times is hard to achieve and needs substantial starting capital as you have to trade quite a number of stocks in parallel in order to achieve the multiplier effect. With the trading opportunities seldom coming at you synchronously, often the potential trades are 'bunched' together so sufficient capital is needed to optimize getting the maximum number of trades available at any one instant.

Our entire methodology has been specifically designed for the individual trader with limited capital. It may be of no great interest to the major brokerages as it does not scale well to trade massive blocks of shares (we need different algos for this purpose).

Using small volume trades coupled with market orders to guarantee instant execution our method seems to hold quite a number of advantages for the individual trader. Our fill rate is virtually 100% as we use market orders almost exclusively and our share volume rarely exceeds 1000 shares per trade. No tier 1 player is likely to be interested to compete in this space.

Recently the high frequency of trading has been validated by a number of workers in this area of finance. This type of high frequency trading should not be confused with the current fad of ultra high frequency trading which makes use of collocation of servers at the Exchanges to enable placing extreme order quantities and also to cancel at an equivalent rate. The ultra low latency approach facilitates machine placement of order rates of over 1000 per second with peaks reported of 8000 per second.

The high resolution methods which we are working with are accessible to the individual trader (in contrast to the huge capital investment required for the ultra-high frequency methods) and will produce a considerable increase in the number of people who will embrace personal trading as a hedge against the market cycles. Also there is no way to describe the self satisfaction that is obtained by controlling one's own finances and making profits based on sound principles, skill and daring (not too much daring . . .).

The results sound so simple to achieve, and to a certain extent they are, but it nevertheless took us 12 years of trading and research, going in and out of a multitude of blind alleys. We spent three years researching Artificial Neural Networks (ANNs) only to find that the technology (2001) could not cope with real time. Moore's Law has given us a hand and we plan to go back over the work we have done to see if we can use the current computer power to use ANNs and possibly other Artificial Intelligence methodology in developing a new type of algo.

To finally reach the level of expertise in the basic methodology described in this book took a fair amount of isolation from friends and family. Trading is both enormously demanding, addictive and intellectually extremely stimulating but also very time consuming. Many a time we started analyzing the day's trading session after the Close at 4pm Eastern only to be surprised that we had spent the night at the computers and that it was 9 am and only half an hour before the Open. It is only too easy to lose all track of time when analyzing the markets.

That the markets and the 'electronic pit' are complex, continuously evolving and ever-changing cannot be stressed enough. It is said that in order of complexity the markets rank a good 4th after the Cosmos, Human Brain, and Human Immune System. Although the latter three are a little bit older and had a bit more time to evolve to the present state.

ANOTHER VOICE

Though the trading paradigms we embrace are fundamentally different to so-called 'high frequency trading' as currently being practiced on Wall Street there is however one major concept which is shared: Minimum holding times and many trades.

Very few practitioners show the freshness of intellect and insight of Irene Aldridge whose article in *FinAlternatives* we have the pleasure of quoting in full. It is rare to read so lucid a dissertation where most others are still stumbling around in the dark. With many thanks to Irene.

HOW PROFITABLE ARE HIGH-FREQUENCY TRADING STRATEGIES?

High-frequency trading has been taking Wall Street by storm. While no institution thoroughly tracks the performance of high-frequency funds, colloquial evidence suggests that the majority of high-frequency managers delivered positive returns in 2008, while 70% of low-frequency practitioners lost money, according to The New York Times.

The discourse on the profitability of high-frequency trading strategies always runs into the question of availability of performance data on returns realized at different frequencies. Hard data on performance of high-frequency strategies is indeed hard to find. Hedge funds successfully running high-frequency strategies tend to shun the public limelight. Others produce data from questionable sources.

Yet, performance at different frequencies can be compared using publicly available data by estimating the maximum potential profitability. Profitability of trading strategies is often measured by Sharpe ratios, a risk-adjusted return metric first proposed by a Nobel Prize winner, William Sharpe. A Sharpe ratio measures return per unit of risk; a Sharpe ratio of 2 means that the average annualized return on the strategy twice exceeds the annualized standard deviation of strategy returns: if the annualized return of a strategy is 12%, the standard deviation of returns is 6%. The Sharpe ratio further implies the distribution of returns: statistically, in 95% of cases, the annual returns are likely to stay within 2 standard deviations from the average. In other words, in any given year, the strategy of Sharpe ratio of 2 and annualized return of 12% is expected to generate returns from 0% to 24% with 95% statistical confidence, or 95% of time.

The maximum possible Sharpe ratio for a given trading frequency is computed as a sample period's average range (High – Low) divided by the sample period's standard deviation of the range, adjusted by square root of the number of observations in a year. Note that high-frequency strategies normally do not carry overnight positions, and, therefore, do not incur the overnight carry cost often proxied by the risk-free rate in Sharpe ratios of longer-term investments.

The table below compares the maximum Sharpe Ratios that could be attained at 10-second, 1-minute, 10-minute, 1-hour and 1-day frequencies in EUR/USD. The results are computed ex-post with perfect 20/20 hindsight on the data for 30 trading days from February 9, 2009 through March 22, 2009. The return is calculated as the maximum return attainable during the observation period within each interval at different frequencies. Thus, the average 10-second return is calculated as the average of ranges (high-low) of EUR/USD prices in all 10-second intervals from February 9, 2009 through March 22, 2009. The standard deviation is then calculated as the standard deviation of all price ranges at a given frequency within the sample.

Time Period	Average Max. Gain (Range) per Period	Range Standard Deviation per Period	Number of observations in the sample period	Maximum Annualized Sharpe Ratio
10 seconds	0.04%	0.01%	2,592,000	5879.8
1 minute	0.06%	0.02%	43,200	1860.1
10 minutes	0.12%	0.09%	4,320	246.4
1 hour	0.30%	0.19%	720	122.13
1 day	1.79%	0.76%	30	37.3

As the table above shows, the maximum profitability of trading strategies measured using Sharpe ratios increases with increases in trading frequencies. From February 9, 2009 through March 22, 2009, the maximum possible annualized Sharpe ratio for EUR/USD trading strategies with daily position rebalancing was 37.3, while EUR/USD trading strategies that held positions for 10 seconds could potentially score Sharpe ratios well over 5,000 (five thousand) mark.

In practice, well-designed and implemented strategies trading at the highest frequencies tend to produce double-digit Sharpe ratios. Real-life Sharpe ratios for well-executed daily strategies tend to fall in the 1–2 range.

By Irene Aldridge – October 2, 2009 in *FinAlternatives*

11
Our Nomenclature

Nomenclature is just a fancy way of saying: 'here are the actual meanings of these common words used with a different nuance.' The aim is to have their precise meaning unambiguously agreed and understood between all the users.

We have found that all successful enterprises and projects develop their own language – lingo develops and acronyms come into common use.

Personally we have found it indispensable even in a team of two to ensure that what we say to each other is understood exactly how we meant it. This precision is not pedantic but in some cases can mean the difference between success and failure. Besides, it can save a lot of time and potential friction. We argue less . . .

We will try to keep our nomenclature consistent, simple and hopefully clearly understandable (perhaps not always using the standard definitions) and hope to succeed in this area by including words which we use in our everyday trading and research work. You would not believe how we have struggled untold hours trying to comprehend what a symbol or an equation really meant in a book or learned paper and we are determined to spare you such agony. So here is the list of definitions and explanations of the key words, symbols and concepts we will use.

Ticks

Tick data = intraday real-time trade data

Tick = trade = one trade, finest resolution in our system

T = trade (we use 'T' as the abbreviation symbol in equations involving ticks)

Number of ticks is expressed with n, thus

nT = number of ticks e.g. 30T, where n = 30

Time

We express time in seconds, with abbreviation symbol 't' thus
t = time unit in seconds
nt = number of seconds e.g. 40t, where n = 40 seconds
Timestamp = the time of execution of the trade as reported
by the Exchange the format is usually hh:mm:ss.

EOD

End of day – data totals for the whole trading session

Backtest

Procedure testing the performance of an algo against historical tick data, over a specified **lookback** period.

TXN

TXN = transaction, usually used in the plural to signify total transactions which took place for a particular stock in a session.

Trade Price

S = Generic trade price symbol displayed as $ to two decimals, $.nn.

Return

Return, assumed $RET usually shown as

$$S_n - S_0$$

S_0 = Designates the Price at trade start
S_n = Designates the Price of the trade at its conclusion.

May be expressed as a percentage by dividing return by Price at the start.

Basis Points

The **return** may also be displayed in 'basis points'

$$\text{bp} = \text{basis point},\ 100\,\text{basis points} = 1\%$$
$$\text{bpRET} = \text{basis point RET} = ((S_n - S_0)/S_0) * 10\,000$$

To convert decimal 0.01 to bp multiply by 10 000 = 100bp

Lookback

Symbol abbreviation = LB, a generic term which is used to specify how far back we look into the historic data in calculating moving averages, volatilities, returns, backtest, EMA α constants etc. This period is chosen to suit the requirements of what data is being calculated and what you are trying to find out.

Maxima, Minima, Ranges

$$\text{MAX S} = \text{maximum trade price over specified lookback period}$$
$$\text{MIN S} = \text{minimum trade price over specified LB}$$
$$\$R = \$\text{Range} = \text{MAX S} - \text{MIN S (over specified LB)}$$
$$\%R = \%\text{Range} = (\text{MAX S} - \text{MIN S})/\text{S AVG (average over LB)}.$$

Average

$$\overline{\text{X}} = \text{average of x series, read 'x bar.'}$$

Volume

$$\text{Vol} = \text{Abbreviation for share volume, usually per transaction}$$

May be quoted for session total.

Moving Averages

$$\text{MA} = \text{moving average}$$
$$\text{SMA} = \text{short moving average}^*$$
$$\text{LMA} = \text{long moving average}$$
$$\text{EMA} = \text{exponential moving average}$$
$$\text{WMA} = \text{weighted moving average}$$
$$\text{MMA} = \text{median moving average}$$
$$\text{AMA} = \text{adaptive moving average.}$$

Parameter

Any user definable constant, used in an algorithmic calculation.

* This is usually reserved for 'SIMPLE MOVING AVERAGE.' Please excuse our usage redefinition but our pervasive use of SMA as a 'short' moving average in all our work would make it even more confusing. We do not use 'simple moving average.'

A Couple of Greek Letters

$\sum$ = Capital sigma is a Greek letter which we use to denote summation,

$=\sum T_0 \ldots T_n$ thus tells us to sum the series $T_0 \ldots T_n$

Sigma $\sum$ is the 'summation operator' and we shall usually use it with no superscripts or subscripts to indicate summation over the entire area under discussion.

Sub- and superscripts will only come into play when we have to define the summed area as a subset.

μ = mean, average, Greek letter mu (pronounced 'me-you'), strictly speaking it is the mean of the 'population,' that is all your data; x bar is the mean of a 'sample' or subset of the population.

Charts

In our charts the 'x' axis is the horizontal axis, and is usually configured in ticks. If we have to go to seconds for equitemporal reasons this is clearly indicated. Otherwise it is always in ticks as the default.

The chart's 'y' axis is always the vertical axis and is in \$. On rare occasions we may use the log of the price when the axis will be in log(x).

Slope

m = slope = $\Delta y / \Delta x$ (The change in y divided by the change in x)

Watchlist

= A list of securities which we want to track, and maintain in a quick and convenient snapshot format. May include various metrics.

Volatility

Please see the Volatility chapter where we will explain the various metrics for quantifying volatility, their meaning and how we use them in our ALPHA ALGOS.

Stats Toolbox

Variance, sigma, standard deviation, average, median, mode, median standard deviation (please see Chapter 13 below).

Math Toolkit

Sigma sign, log, ln (natural logarithm), exponents, first and second derivative (please see Chapter 12 below).

12
Math Toolkit

Our math toolkit is intentionally quite sparse as we have seen that smothering the trading problem with complex math did not provide us with commensurate benefits in the way of returns. Most algo construction can be achieved with the simplest of tools. We think that this is the most you will need to start with.

One of the most important concepts for us is **Moving Averages**. For the strategy of ALPHA ALGO design moving averages of the price, volatility, return density, and intertrade duration are the key ingredients to the recipe. In this book we will be concentrating only on stock price as the input data, leaving the analysis of the other factors for future volumes.

We shall now describe these moving averages in some detail as they constitute the foundation stone and raw material building blocks of most algorithmic strategies from the most basic to the ultra advanced.

THE ARITHMETIC MOVING AVERAGE

For the SMA, (Short Moving Average) and LMA (Long Moving Average) we use the plain vanilla arithmetic versions. (NOTE that in our system 'SMA' stands for Short Moving Average and not Simple Moving Average as is normally the standard usage.)

The basic concept of these moving averages can best be visualized by imagining a 'window' of fixed size and sliding it over the number series, as new data appears at one end and old data is deleted so the total items in the window stay constant. Averaging these numbers at every step (every time we move the window one step) gives us the 'moving average' series.

Now please go to your computer and open the file 'MOVING AVERAGES' on your CD and go to sheet tab SMA. This example will demonstrate how moving averages work in general. The jagged blue line represents the trade price.

We usually color code the SMA line white on all charts. The 'window' or lookback on this plot of the SMA is 100T (100 ticks). Before we move to the next tab here is a useful shortcut when you are experimenting to find a good short moving average: Place your cursor touching the blue price line and right click – this will pull up a window where the MENU bar is 'Format Data Series.' Now scroll down to 'Add Trendline' and select it. This will pull up the 'Add Trendline' window. In the second row, third on the right you will find 'Moving Average.' Select it. (It will turn black, and so will the little length scroller.)

You can now scroll up to a maximum of 255. This maximum is a bit limiting but for quick short MAs it is very fast and convenient. It's extremely handy for testing and checking things quickly. We use it constantly to quickly see if there are any short patterns we may have missed.

Moving on to the next tab, LMA, this plot is usually color coded red. Here the window is longer, the lookback = 300 T.

Now move to the next tab, SMA & LMA where all the plots are shown together on the same chart.

Notice that the blue line, which is the trade price tick line, is jagged. The white line is smoother than the blue line, many of the 'jags' having been smoothed out, and the red line is even smoother than the white line.

So we see that the length of the lookback period is what primarily controls the 'smoothness' of the line. The longer the lookback, the smoother the line. Unfortunately the longer the lookback the less 'up to date' with the latest real-time value is the current average, as we have introduced a certain amount of lag into the system.

The formula for the MA is:

$$\mathrm{MA} = (\Sigma \mathrm{T}_0 \ldots\ldots \mathrm{T}_n)/(n + 1)$$

where 'n' is the lookback period.

This whole area really falls under the rubric of DSP, Digital Signal Processing, and we shall touch upon how we shall use 'convolution' in the ALPHA ALGOS.

'Convolution' is a formal mathematical operation just like multiplication, addition and integration. Where addition takes two numbers and produces a third number, convolution takes two signals (we are considering our tick series as signals) and produces a third.

Please move to the tab on the CD marked 'TRIGGERCONVO.' Here we have a third line (in green). This is the result we obtain when we subtract the LMA from the SMA. Note that the values are much smaller and have to be displayed on a different scale, on the right hand axis. (Please see the Excel Mini Seminar chapter on how to do this.)

This is, in fact an early introduction to our Trade trigger signal line in our ALPHA-1 algo which you will be seeing later on. This particular trigger line has not been parameterized so is only shown here as an example of how moving averages are constructed and how convolution works. You can scroll right to see an expanded version of the chart.

Another useful concept to keep in mind is that moving averages have 'filter' properties. Long moving averages are 'low-pass' filters whereas short moving averages are 'high-pass' filters. A 'low-pass' filter attenuates (damps, reduces) the high frequencies while letting the low frequencies pass through unimpeded. A 'high-pass' filter attenuates the low frequencies and lets the high frequencies pass through.

The field of Digital Signal Processing is most useful in the design of trading algos but beyond the scope of this volume. If you are interested in this subject we found the most approachable text-book to be by Steve W. Smith – please see the Bibliography.

There are quite a number of other moving averages, such as median, weighted, logarithmic, standard deviation and exponential (EMA) as well as a number of others. In fact any variable can be turned into a time series homogenous or inhomogenous and charted to see how it unfolds.

We should not leave out the area of 'transforms' where we carry out some mathematical operation on a series in order to obtain some specific effect – such as for example compressing the range of the series.

The EMA is of particular importance to us as we find that much of the time it gives some of the best results.

EXPONENTIAL MOVING AVERAGE, EMA

The formula looks a bit frightening but it is really quite easy to do in Excel (please have a look at the file 'EMA' on the CD).

$$EMA = EMA_{t-1} + ((2/(n+1)) * (S_t - EMA_{t-1}))$$

where

n = the length of the lookback
S = trade price.

Note that the first EMA value is seeded with the earliest trade price (oldest trade price).

EMA_{t-1} is the prior EMA's value.

With a simple MA, all the data has an equal weight in the calculation of the average with the earliest data dropping away from the MA 'window' as new values come in.

In the EMA the earliest (oldest) data is never removed, but as you move the series forward its influence diminishes according to an exponential decay function which you have set by selecting the length of the lookback, n.

The 'speed' of this decay function is determined by the length of the lookback. Thus the earliest trade prices have the least influence on the current value of the

EMA. In our charts we color the plain vanilla EMA a bright violet (or any visible color on the chart background you have chosen).

Here are some more odd items of information which will come in handy understanding some of the nuances.

Deviation from the mean is usually expressed as 'absolute,' which means always positive. The sum is without sign and 'absolute' is usually indicated by two vertical bars encompassing the expression

In some cases a formula is used for determining the next term of a sequence from one or more of the preceding terms. This is called '**Recursion.**' The function is 'recursive.' EMA series are recursive.

In computer language a 'for loop' is a statement which allows the target code to be run repeatedly. This type of loop is classified as an '**Iteration**' statement. Iterative means repeating.

You are probably quite familiar with the next few sections but we have thrown them in for completeness for those who had the good fortune to miss the classes covering these items.

One thing to remember is that variables are all the things that vary and constants are all those values which do not. Parameters are constants where you set the values.

EXPONENTS AND LOGS

Understanding how **Exponents** work comes in handy when you will analyze **Returns**.

X^2 means X times X. The little 2 superscript is the exponent. It can be fractional, for example $X^{\frac{1}{2}}$ means the same as $\sqrt{X}$, or square root of X. An exponent can take any real value e.g. 0.2, 3, 5.5.

LOG to base 10 are usually written log(x) and natural logs 'ln' to base 'e' usually written ln(x). Most of our math uses the natural log. Both work the same way.

If you remember the basic log operations well and good, if not, here is a quick review.

A logarithm is the number to which you must raise your base (10 for ordinary logarithms) to get your target number. Thus $\log 100 = 10$ (because you raise the base, 10, by a power of 10, thus $10^{10} = 100$).

Logs have some useful properties.

They change multiplication into addition: $\log(nx) = \log n + \log x$

Logs change division into subtraction: $\log(n/x) = \log n - \log x$

Logs change raising to a power into multiplication: $\log(10^{10}) = 10 * \log 10$.

Trade prices are sometimes expressed as the log(trade price) especially when the graph axis has to take a large range.

There is also the advantage that plotting logs of an exponential series will produce a straight line plot. Many practitioners use the natural log (logs to the base 2) return as the default.

CURVES PLUS

No, not what you were thinking but just as useful as all series can be plotted as curves on a chart. This is our main point of reference and understanding of how the trading properties develop and how we can understand and assess, extract meaning and make decisions.

All curves have an underlying equation from which they are generated. It is useful to get a mini view of how equations translate into charts.

The parabola is a beautiful curve that you will be seeing a lot of. Its equation is $y = x^2$.

Please see the file 'CURVES' on the CD for examples of parabolic curves in their pure form with their generating equations. The terms 'curves,' 'lines' and 'plots' are equivalent.

Power Law Equations have the form of: $Y = ax^b$ where a and b are constants – this is called a power law as Y changes according to the power X is raised to.

Nonlinear Exponential Equations can have the form of: $Y = \alpha e \beta X$ where α and β are constants and e =2.71828 and X is the driving exponent.

Obviously things quickly get complex when various equations interact and convolve. What we observe in the price trajectory is this ultra-convolution of all the factors which influence the price of the stock, a cocktail of equations and parameters, continuously in a state of flux.

FUNCTIONAL NOTATION

Functional notation is useful shorthand: f(x)=2x, read 'f of x equals 2x.' Thus the output of the function is dependent on the value of x. Of course the expression following the equal sign can be complex but it will always be evaluated by what the x value is.

SLOPES

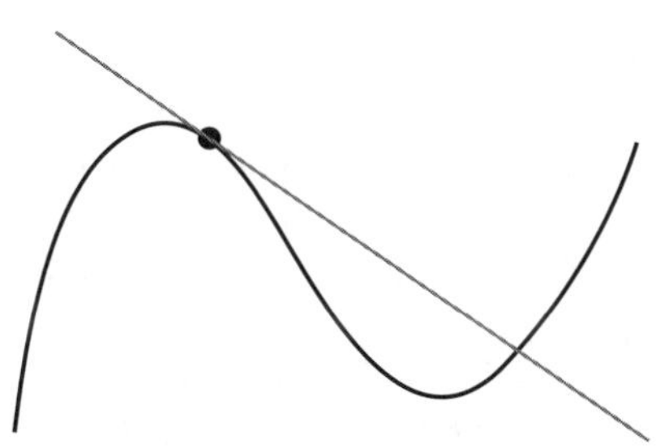

Tangent Line Function Line

The **Tangent line**, the straight line, is drawn to the curved **Function line**. The slope of the Tangent line is equal to the derivative of the function at the point where the straight line touches the plot of the function line.

For the **Slope** calc we have a sort of mnemonic – imagine a right angled triangle with the hypotenuse touching the function line.

The slope is then 'opposite' triangle side divided by 'adjacent' side. (The scales must be the same, unless you use it strictly as a standard for comparisons.)

Or you can use $\Delta y / \Delta x$ [read delta y divide by delta x].

To recap: The slope is the vertical distance divided by the horizontal distance between any two points on the line. In all cases the delta value must be kept small.

DERIVATIVES

My dear calculus professor would take issue with this definition but it works for what we have to do with it: The **Derivative** evaluates how a function changes as its input changes. (In calculus these changes are at the limit of being smallest, at infinity.) It can be thought of as how much the value is changing at a given point. It is the best linear approximation of the function at the point of contact.

The derivative of a function at a chosen input value on the plot describes the best linear approximation of the function of that input value.

Thus the derivative at a point on the function line equals the **Slope** of the **Tangent line** to the **Plotted line** at that point.

There are many notation conventions – Leibnitz, Lagrange, Euler and Newton all have their own way of writing the derivatives.

We shall use that of Leibnitz.

As an example, for the equation $y = f(x)$ the first derivative is denoted by dy/dx.

An example of the first derivative is the instantaneous speed, miles per hour. The instantaneous acceleration is given by the second derivative.

SETS

The theory of sets is a branch of mathematics. Sets are collections of defined objects.

George Cantor and Friedrich Dedekind started the study of sets circa mid-1870s. Surprisingly the elementary facts about sets are easily understood and can be introduced in primary school, so that set membership, the elementary operations of union and intersection as well as Venn diagrams are used in the study of commonplace objects.

Any 'object collection' is amenable to the precepts of set theory. Sets can have subsets contained in them and equally there are supersets which can contain other sets.

We shall not go into any detail of set theory but we found the idea that we can define any collection (just as we may like to define it) of things, ideas, people, stocks as a 'set' to be most freeing for the imagination and to fan the creative fires.

We find thinking in sets an extremely flexible tool – like 'this set lists all the stocks we will trade priced at $70 and above'; or 'the set of stocks having a % Range over N are easier to trade with ALPHA-1.'

Here is an example of how sets come in handy: We have two hypothetical lists of stocks – List 1 is filtered by price: more than $30 and less than $100. List two is filtered on daily volume being not less than 800 000 shares.

Let's look at List 1 (Set 1) and see how many of these stocks have traded above the Set 2 volume 800 000 filter. This 'subset' can now be looked at for properties which might lead us to a trading strategy which we can design into a trading algo.

The concepts of **Sets** and **Clusters** are related and we find them both useful in the selection of stocks to trade. Please see the Bibliography for further reading.

13
Statistics Toolbox

This chapter is intended to give you the tools needed to get a better understanding of the characteristics and properties of the data streams you will be looking at.

There are bookshelves of statistical books which we found too wide of reach for the analysis we had in mind. What we need are the tools to express our concepts, many of which are visual impressions, in a melange of mathematical terms, as simply and concisely as possible.

We use statistics to describe and give us a picture of the characteristics and properties of the market data with which we work to create ALPHA ALGOS. Both the entire populations as well as samples taken from them are used.

First of all we must look to see how much data we have – is it the whole 'population,' or do we only have a 'sample' (or samples) from it? As we already know a sample is a subset of a population.

We assume that the data is 'clean' with errors and outliers already washed out by the vendor. This depends on the data vendor selection. Tick data tends to be cleaner than aggregated data. In our experience this has never been a problem. If it does happen it is usually so far out that any of our algos would ignore it. Should it become a problem we can filter the input data but so far we have never had the need to do this.

Next we would like to know something about how the data is distributed. Is it tightly clustered close together, or is it strewn all over the place? Are there places where it is more concentrated or places where it varies more than it does in others?

We use subscripts as position markers, e.g. T_{10} means that this is the 10th tick. We define the return intervals using the subscripts. So from first trade to 12th trade the return interval is $T_{12} - T_0$ is 13 ticks,

To signify a number of ticks we use **nT**. Thus a lookback of 150 ticks will be written 150T.

The following tools can be used on any data. If we have the series we can use any part we are interested in by defining a lookback length and operating on that.

These tools are all demonstrated on the CD filename 'CENTRALITY.'

A useful trick if you want to make the function formula of a cell visible is to precede it with an apostrophe, like so ' – this turns the contents into text and displays it. Also, it turns back to a function if you remove the apostrophe.

The **mean** or **average** (they 'mean' the same thing in our usage) is a rough first look at the data we are going to analyze.

The sample average is the sum of the variables divided by the number of data points.

$$\bar{S} = \text{average price} = \Sigma s_i/n$$

$\bar{S}$ (often read as S bar) is the dollar average over the lookback number of ticks, n.

The mean is a most useful function and much can be learned by using different lookbacks, comparing them, and looking for sudden changes to hunt for telltale prodrome signs of trend change.

The average deviation or mean deviation is a useful measure of how much the data swings around the 'center.'

$$\text{MD} = \text{mean deviation} = \Sigma|S_i - S/n|$$

The two vertical bars mean 'absolute,' negative signs are read as plus in the result of the expression within the bars. This can be quite confusing till you get the hang of it – that the minus sign when between two vertical bars is read as plus sign.

We also found it useful sometimes to look at the +DEVIATION and the parallel negative DEVIATION over short tick lookbacks and see how much they vary. The total negative here is then made positive to compare with the +DEVIATION.

If you wish to know how often a particular value appears in the series you choose the **mode**.

The **median** is the value in the middle of a sorted series lookback. The median has an advantage over the average in that it is less influenced by 'outliers,' that is values far from the center. It is sometimes called the '50th percentile' because half the data is below the median. It is also sometimes called the second 'quartile.'

The first quartile is the 25th percentile and the third quartile is the 75th percentile. These are what we call 'measures of relative standing.'

The percentile is the value of a variable below which a certain percentage of values fall. The 10th percentile is the value of the variable below which 90% of the other values fall.

THE Z-SCORE

Another very useful measure of relative standing is the z-score. This is the distance that a value lies above or below the mean of the data measured in units of standard deviation, sigma. More on standard deviations later in this chapter.

$$\text{Thus the z-score} = (x - \bar{x})/\sigma$$

A negative z-score tells us that the value x lies to the left of the mean, while a positive z-score tells us that it lies to the right of the mean.

The problem with outliers (values which appear not to be correct and not to belong to the series) is that it is always quite difficult to establish with any degree of certainty whether a value far from the average is a 'true' outlier or not. It could be a genuine characteristic of the particular distribution you are analyzing or it could be some anomaly such as a mistyped trade that got through or a transmission error, or even a blip on the line due to some mechanical or electrical disturbance.

Outliers may be found using z-scores and Tchebysheff's Theorem tells us that almost all observations in the data will have z-scores of 3 or less (please see below). Thus a much higher z-score would indicate that the value is an outlier and should be ignored in the main deliberation on the dataset.

To visually compare the relationship between two quantitative variables we can use the Excel scatter plot.

We can also use the Excel Side by Side column graph feature to look at how two variables relate to each other visually.

To get a visual handle on how the data is distributed we construct a **histogram**. Note that we are not making any assumptions on how the data is distributed – we are just dispassionately looking to see what we have.

Excel is not the best histogram maker but will do for most work. Please look at the File: 'HISTOGRAMS' on the CD (here we have used www.treeplan.com 'Better Histogram' software) which illustrates some of the points below.

The histogram is a most useful tool for 'reality testing.' You can quickly assess the maximum number of returns at a certain value and tick length.

Place the tip of your cursor on a histogram bar and up will come a little window with the message:

Series 1 Point "13.46"
Value : 93

The 'Series 1 Point' designates the value of the return in basis points (13.46) while the 'Value' tells us the number of returns with that basis point value in the series.

This gives us a rough idea of what it is possible to achieve with this stock under the current market conditions

There are a multitude of 'distributions' which can be constructed from their underlying equations. We fundamentally make no assumptions that the data belong to any particular distribution.

We must be aware that the actual price trajectory is a convolution of an unknown number of factors (equations) which in themselves fluctuate and also fluctuate in their level of contribution to the total convolution resulting in the trade price.

The most well known distribution in statistics is the so-called 'Bell curve' which works perfectly well in a multitude of cases. Many items in nature follow this distribution to the point that it was called the 'normal' curve. It has been the workhorse of statisticians for well over a century.

Karl Friedrich Gauss (1777–1885) was one of the giants of mathematics. He and Abraham Moivre are credited with the discovery of the 'Bell' curve. Both published results in 1733. Gauss actually derived the equation for this distribution. Pierre-Simon Laplace was also very involved in the popularization of the 'normal' distribution.

This distribution has been a point of contention in finance for a very long time – with the '**Gaussian**' distribution initially being the most used. Professor Mandelbrot in his study of early cotton prices found that this distribution did not reflect the entire truth of the data and that it had exaggerated legs, fat legs. He proposed a more refined distribution but this has also only been a way station as research has still to come up with a definitive answer as to how financial asset series are really distributed.

However, the Gaussian distribution works up to a point (just shut your eyes to the simplifying assumptions for the moment) and in addition it has a bevy of analytic tools which have developed around it which we can use as long as we are aware that the assumptions we are using may turn around and bite us (on occasion . . . like when a market crashes and hedges do not hedge).

Most of the computations in the past required that the data be IID, meaning that each measurement was taken independent of other measurements and that all measurements came from an identical distribution, thus IID. This is a stringent requirement and financial data is hardly ever that compliant. So whenever you come across the innocent looking IID, watch out! It's not what your data is likely to look like and any computations using this constraint are likely to be built on a foundation of sand (paraphrasing Professor Mandelbrot yet again).

In 'high resolution land' it is best not to make any assumptions as to the data distribution but to look at the actual tick streams and go from there. What we are really after is some idea of the dynamic trajectory of the price and this is more in the realm of digital signal processing.

A useful tool is the VARIANCE, s^2, which is equal to the sum of the squares of the deviations from the mean.

We interpret the sample variance as a measure of how spread out the data is from a center.

STANDARD DEVIATION

Probably the most used measure is the **standard deviation**. This is the square root of the variance. It is a good measure of dispersion for most occasions and worth the fact that we are using it in violation of some of the precepts on how it operates. We suspect that one reason for its popularity may be that it is so easy to calculate ...

The symbol we use for the standard deviation is the Greek lower case letter, sigma, σ. In a Gaussian distribution we can usually get 95% of the values within $\pm 2\sigma$, and approach 100% at $\pm 3\sigma$.

The standard deviation is the usual way we measure the degree of dispersion of the data, Gaussianity is assumed with clenched white knuckles and crossed fingers.

The formula for actually calculating this distribution from first principles is a bit complex so we satisfy ourselves with getting the standard deviation using the Excel function STDEV.

=STDEV(A1:A200) gives us the sample standard deviation of the series in Column A Rows 1 to 200.

$$\sigma = \sqrt{\Sigma(x - \bar{x})^2/(n - 1)}$$

Please look at the 'CENTRALITY' file on the CD.

We have made little differentiation between 'sample' and 'population.' A sample is data drawn from a population at random. The 'population' is the entire dataset.

'Stratified' samples are obtained after you have divided the population into a number of distinct and non-overlapping strata.

CORRELATION

'Correlation' (R) is a measure of association between two variables.

In Excel the function 'COREL' will calculate the value of 'R' which is unitless, and can take the values of 1 to 0 to –1, where 0 means there is no correlation, 1 stands for complete positive correlation and –1 stands for complete negative correlation.

In all cases the variables must be 'paired' and each series must have the same 'length.'

If you build a scatter plot of the two variables and you get a straight line rising from left to right and the points cluster strongly about the line, the relationship is positive, R will be close to 1. If the line falls from left to right the relationship is negative and R will be close to –1.

Many financial series exhibit the property we call 'autocorrelation,' which means that following values are somehow dependent and correlated to earlier values. This feature negates most of the classic statistics as the proviso there is that the values be

IID ('independent and from an identical distribution,' remember?). In most cases the autocorrelation lasts only a relatively short time, say a maximum of five minutes, and then vanishes.

However in a high frequency short holding time trading horizon this we believe is of the essence. It is just the stylistic anomaly of the market which we can use to make profitable trades.

Many strategies for creating powerful trading algorithms may require that the incoming data be manipulated in some way. For example we may want to remove the high frequency element in the data series – this we would do with a smoothing moving average – which is essentially a 'low-pass' filter. It lets through the low frequencies while attenuating (curtailing) the high frequencies.

To recap: A short moving average is equivalent to a 'high-pass' filter. It attenuates the low frequencies leaving us with the high frequencies. In other cases we may need to remove the low frequencies and keep the high frequencies. This trick may also be accomplished by 'differencing' successive values of the series.

Differencing is a data transformation which we make by serially and systematically subtracting one value from another through the entire series of the data with each pair of series values separated by a 'lag.' The lag minimum is one step but a larger lag may be required to achieve the desired 'high-pass' filter.

'First differencing' is the preferred filter we use to detrend the data – the high frequencies (and unfortunately any high frequency noise) remain. Please refer to TRANSFORMS on the CD.

Another useful maneuver is to 'standardize' the data series. This is accomplished by subtracting the mean from all observations and dividing by their standard deviation. The series is then deemed 'standardized' and is easily compared with other standardized series.

14
Data – Symbol, Date, Timestamp, Volume, Price

In our present methodology we have restricted the input data quite severely, taking in only: SYMBOL, DATE, TIME OF TRADE, VOLUME, PRICE.

These are the five column headings which take the real-time data which we use in our ALPHA ALGOS. We often only use Price and Symbol.

Our input data is either provided bundled from our brokerage or from a specialist real-time tick resolution data vendor. Please have a look at the chapters on Data Feed Vendors and Brokerages in due course, as you will have to decide on the suppliers to bring your tick real-time data feed into your spreadsheets. We will leave this area for the relevant chapters later in the book.

For each stock on our Watchlist (either from the NASDAQ or NYSE Exchanges) the data feed provides on request realtime trade data, as it comes from the Exchanges (say via your brokerage account) on the headings which the feed handler writes into our Excel templates, one stock per template. As previously mentioned, each template is running under its own Excel instance.

The following descriptions may take a bit of following but if you look at the CD file 'DATA' things will become a bit clearer.

The ticker symbol of the stock is written into Column A Row 1. Excel defines the columns in the spreadsheet either by number starting at 1 (=Col A) or alphabetically with A being the first column.

As already mentioned before we shall use alphabetic Column designation throughout this book.

Into Col A Row 3 of the Excel instance goes the Date of the trading session formatted as mm dd yyyy.

Into Col B Row 3 goes the Timestamp of the trade formatted as hh mm ss. Some vendors have the capability of providing millisecond resolution but we have not found it necessary up to this point. Perhaps as the markets evolve this resolution might become useful but at present we do not find it necessary.

Into Col C Row 3 goes the Volume of the trade, in integers. This is the number of shares traded on this transaction. (This is the number that is shown in the Time and Sales column of your Level II display.)

Into Col D Row 3 goes the Trade Price in $, with two decimals for the cent part, like so 23.89. Some feeds do not write the double zeros for whole dollar values.

(The dollar value is also shown on your Time and Sales scrolling column of your Level II display.)

In order to clean up the headers which contain the request strings to the data vendor's server we copy Cols B, C and D to Cols E, F, G and H (all into Row 3). This not only gets rid of the calling strings but also provides us with freedom to move the data around. (The initial 4 columns are written as an array in Excel which does not allow any flexibility as arrays cannot be edited.)

The above is standard for all our templates.

In the example on the CD we have added Cols G and H which contain SMA 200T and LMA 600T respectively. We have also added in a DIFF column 'I' in which we subtract the LMA from the SMA to obtain the DIFF TRIGGER. Just a taste of what is to come.

15
Excel Mini Seminar

In keeping with our aim to keep this book as 'free standing' as possible we will run through some of the essential Excel stuff that is required for the design of algorithms and implementation of algorithmic trading. We have chosen Excel to implement our algos as it is the universal spreadsheet with the added advantage of a fairly rich function language in optimized microcode and thus provides fast execution.

We will restrict our rambling through the Excel forest strictly (or as close to strictly as we can manage without getting lost) to what we will be using in our algos.

We have lived with spreadsheets since their very inception and the industry standard **Excel** has been the work horse for nearly all of our research.

Please have look at the file 'EXCEL MINI SEMINAR' on the CD.

We are using Excel version 2003 rather than 2007 (which is the latest version) except where clearly stated (the reason is that 2007 has much expanded grid size with one million Rows, where 2003 only gives us 64K Rows and 256 Columns. This can become a problem with high activity stocks as each transaction takes up one row.)

Every 'instance' of Excel is the complete application. We use one instance for each stock that is traded. This provides us with a safe, redundant, well behaved and very fast overall system. Obviously we 'pay' for this extravagance by having to use much faster multiple processors with large amounts of RAM and some very fast hard drives, preferably the new solid state type.

Every instance of Excel can have further sheets added to the default 3. We prefer not to go over 12 as that seems to be our limit for keeping oriented.

When possible it is good practice to keep the Excel file size below 20Mb, certainly below 30Mb. This decreases the chances of Excel or the operating system misbehaving. We must confess that we often use quite huge Excel files, 125Mb plus. This is not recommended. Do it only if you absolutely have to and be prepared for 'unusual' antics. Use multiple backups.

You are probably an expert Excel jockey, if so, please skip the following – if you are at all unsure give it a go.

So let's get started. Pull up the file 'EXCEL MINI SEMINAR' on your computer and let's go. The workspace has been set up and configured to fit with the work we shall be doing on the algo templates and in general is the same throughout except when an exception is flagged. Each Excel 'book' contains a number of sheets, if you look at the bottom of the Excel page you will see a tab named 'Sheet1' with greyed Sheet2 and Sheet3. If you would like to add sheets select 'Insert' from the top Menu bar then scroll down to 'worksheet' and left click and lo and behold you have Sheet4.

For ease and speed we will describe actions like the one you have just done in a sort of shorthand: On the menu bar click in sequence: `Insert|worksheet`

This shorthand is also useful when designing and note keeping.

To get the sheets in the order of your choice just put your cursor on the tab named Sheet4, click, hold down the cursor and drag it over past Sheet3 and let go when you have reached the position you want, just past Sheet3.

To save your work (without naming the file) at any time just press Ctrl+S. This is a useful habit to develop. Frequent saves will do much to eliminate or perhaps we should say minimize the chance of suddenly finding the system crashed and losing your work. It's totally soul destroying if you have finally solved the code for an algo to have it vanish in front of your eyes. So plenty of saves will avert this problem. (It is also a good idea to have a 'flash drive' to which important files can be backed up as required.) There is also a handy safety feature which saves a copy automatically on a timescale which you define. You will find it at `Tools|Options|Save`. If you don't want it you can disable it.

To name the file go again to the Menu Bar:

`File|Save As|Type in your filename|Save`

Practice navigating around the workspace area using the up, down, right and left arrows. Try PgDn and PgUp. Now use your glidepad or mouse to move the outline cross cursor to place the 'active cell' anywhere you like (it will show it's active by the rectangular bold outline). Type in a couple of numbers – see, it works fine. Now just hit the Del key to clear the cell.

Each cell location is defined by a letter and a number. The letter refers to the Column and the number refers to the Row.

We shall use letters for the Column identifier in all cases.

(Just for your information you can configure Excel so that the columns are also using numbers but for the start that is unnecessarily confusing and we shall use letters to identify Columns throughout this book.)

The Excel workspace continues on off your screen to the right. Select a cell and press the right arrow and hold it down – you will notice that the sheet has scrolled to the right and stopped at Column IV (that is letter I and letter V). Now for the high speed trick: Ctrl + left arrow and you are whizzed straight back to Column A.

You can also scroll (perhaps I should say 'page,' as it moves in blocks) the worksheet using PgDn+Alt to the right and then PgUp+Alt to the left.

Important key shortcuts are Ctrl+C which copies the active cell (or as much as you have highlighted by holding down Shift and using the arrows to highlight what you want to copy – this is called a 'cell range'). Cell ranges are defined by the top cell and the bottom cell, for example (A1:A200) defines the range from Column A Row 1 to Column A Row 200.

Cell ranges of more than one column are referred to like this example, (A1:B200) means the range from A1 all the way down to B200, thus B1 to B200 is also highlighted.

Next you want to 'paste' what you have 'copied' so locate your active cell where you want the stuff to go and Ctrl+V and it's all done.

You can highlight a whole column by clicking on the Column letter in the header. Similarly highlight a whole row by clicking on the Row number.

Now for some calculations, which is what spreadsheets are all about. To flag that you want a calculation in a cell begin with the equal sign, ' = .' That tells Excel that a calculation is coming.

You can activate any cell by placing the cursor (hollow cross usually) in the cell and left clicking.

So let's go to locate the active cell in A3 and type =A1 + A2. This will add the values in cell A1 and cell A2 into cell A3.

Similarly we can use the other arithmetic operators – multiplication is signified by the asterisk ' *,' and division by ' / ' while raising a number to a power is ' ^ .'

When you want to make an entry in an active cell just type into it. What you type will also appear in the edit box of the formula bar, where you can edit it as well, just as you can in the active cell.

In case you want to change the worksheet name just double click on the sheet tab and type in your new name.

If you want to undo a highlight hit 'Esc.'

Here are some bits which are not all that intuitive: Copying formulas can be tricky.

The reason is that Excel uses 'relative addressing' (which you will soon find to be an indispensably powerful feature) so that when you move a formula to a new location Excel will change the cell references to reflect the new location. For example let's say that cell C1 contains the formula =A1 + B1. If we copy C1 to C2 the formula will magically change to =A2 + B2.

This is a crucial feature of Excel. It gives you the capability to copy down a formula for an entire column and still have all the relative addresses correct.

However, sometimes you do not want this relative addressing and want it 'absolute,' unchanging when you copy things around. To achieve this Excel demands you insert a $ sign in front of what you want to remain absolute.

This can be either the Column ref or the Row ref or both (mixed is OK but can be pretty confusing so be careful). So $A$1 will contain whatever is in Column A Row 1 and wherever you enter the absolute $A$1 this value will be copied intact.

Here is a trick: if you want a formula to show as text in the spreadsheet as a marker for yourself of what you are up to, just put an apostrophe in front of the equal sign, e.g. '=A1+A2. This will display your formula, including the apostrophe at the front. You can later remove the apostrophe and it will revert to being a formula.

Formatting the look of the sheet is standard Office stuff.

We find the cell coloring crayon useful to keep oriented or to 'bookmark' something temporarily.

The Bold button (select what you want emboldened and hit 'B' on the formatting toolbar) can also be used for whole columns or rows. Ctrl+B does the same thing.

The format paintbrush which lets you copy formats from one place to another comes in handy as well. If you need to transfer format to more than one place double click when you 'pick up' the brush. Careful, it can be tricky. Centering a column's numbers can be useful.

Charts in Excel 2003 have a special chapter devoted to them a little later on.

Excel has a very rich function language. The added beauty of it is that the functions are optimized in microcode so execution is blazing fast. They are of the form =sum(A1:A10).

You will encounter these a lot in moving averages. Please open the file 'EXCEL FUNCTIONS' on the CD.

An important Excel feature is the ability to test for equality using the 'IF' statement. It works like this: Let's say we locate our active cell in C1 and enter =IF (A1 = 10,1,""). This reads: if cell A1 is equal to 10, then write '1' in C1 cell, if not, write 'nothing' (which is denoted by double quotation marks).

Boolean logic is an indispensably useful tool in devising algorithmic strategies. George Boole was hugely gifted with the ability to learn very rapidly. He studied the works of Newton, Laplace, Lagrange and Leibnitz and created a logic language (around the 1850s) which we now call Boolean algebra. It is the foundation stone of all logic operations and electronic circuit, software and chip design.

Excel supports all the Boolean operators (always write them in capitals): AND, OR, NOT. These are 'bitwise' operators and can only have two values: TRUE or FALSE, 1 or 0.

You can thus test for equal, greater or less than. If the test is successful (Boolean TRUE) you can use '1' as your marker. Using AND, OR, NOT you can put together a fair amount of logic for your algos to work with. You can describe complex logic using a 'TRUTH TABLE.'

A condensed truth table notation:

∧	F	T
F	F	F
T	F	T

V	F	T
F	F	T
T	T	T

Any number of operators may be chained. ANDs, NOTs and ORs can be chained and nested (but the mixture can be confusing so we advise not to use it, unless absolutely necessary – better get the logic to work some other way. . .). Parentheses help, with the operators in the innermost pair being done first, followed by the next pair, till all parenthesized operations have been accomplished. You can then take care of any operations which have been left outside the parentheses. Excel abides by all these rules.

We find it is easiest to work from the left-most column over to the right, and from the top down. This is also how Excel recalculates so we are being kind to the machine. We often split up an algo into a number of columns to give us the transparency that we are doing what we think we are doing. This can later be compacted if necessary. There is not much in the way of execution speed improvement.

Some of the more important Excel functions (preferably always capitalize them) are:

STDEV – Gives you're the standard deviation of a range, thus = STDEV(A1:A1200) will give us the standard deviation, σ, of the contents of the range A1 to A1200.
SUM – is self-explanatory, =SUM(A1:A50)
AVERAGE – this too
MEDIAN – this provides us with the middle value of the specified range (here Excel has to hustle a bit because it must first sort the range in ascending or descending order. . .)
MAX – gives the maximum value of the specified range
MIN – gives the minimum values of the specified range
SLOPE – Returns the slope of the linear regression line through data points in known y's and known x's. The slope is the vertical distance divided by the horizontal distance between any two points on the line, which is the rate of change along the regression line, or as we have previously mentioned it the 'opposite' over the 'adjacent' of the triangle whose hypotenuse touches the curve where we want to evaluate the slope.

In creating algo strategies and writing them in function code we will come upon a number of other functions which we will explain at the point of use. Browsing the HELP 'functions' is a useful exercise.

Moving a selection: Pretty easy, highlight, CTRL-X and the highlighted range will temporarily disappear. Now place you cursor where you want this to reappear and CTRL-V and hey presto, there it is.

The 'HELP' file in Excel leaves a lot to be desired but will do most of the time.

A method we have seen used is to query Google with the question about Excel and it is usually successful in providing a useful answer.

16
Excel Charts: How to Read Them and How to Build Them

The knowledge of handling charts in Excel easily transfers to handling charts in your Order Management System and to specialist charting programs.

We are using Excel version 2003 throughout this book and advise you do the same unless forced to use version 2007 for capacity reasons.

Excel charts limit plots to 32 000 datapoints per chart, and safely up to six plots per chart. More than six may cause problems.

As we have already stressed many times the visual side of our research is key to most of our analyses so proficiency in the use of charts is an important skill.

Open the file on the CD named 'CHARTS.' The ALPHA-1 chart shows what this algo looks like in the flesh.

We are going to go into some considerable detail in how to set up, modify and generally use charts in Excel as this is one of the most important skills which make the creation of algorithms, from the very basic to the very advanced, immeasurably faster and easier to deal with.

It's going to use baby steps but if you follow them all the way through you will have gained a crucial tool for algo development and price movement analysis.

A good facility for creating and working with charts reduces the inertia one sometimes feels when facing a testing task like: 'How does the %Range vary on 150T blocks in a session?'

Once the skill is learned it's almost automatic.

If you want to create a chart here is what you do:

Highlight the data range you want to chart.

Go to the main menu and hit the chart icon.

This will bring up the Chart Wizard panel.

In the 'Chart type' window use the down arrow to scroll to 'line.' 'Chart sub-type' default is the sub-type 'Line with markers displayed at each data value.' If you fancy a preview of what the chart will look like hit the bar titled 'Press and Hold to View Sample.'

Hit 'Next' which will bring up a panel titled 'Chart Wizard – Step 2 of 4 – Chart Source Data.' This will show the chart you are producing in a slightly larger format.

Data Range should show the formula for the range you have selected to chart, e.g. =Sheet1!$AH$7:$AH$15. This is the address of the range you are charting. Note the $ signs which make the address 'absolute' (fixed).

In the 'Series in' make sure the radio button for 'Columns' is selected.

Hit 'Next' – this will bring up the panel titled: 'Step 3 of 4 – Chart Options.' (You will notice that on the bottom of the panel there is a navigation strip with 'Cancel,' 'Back,' 'Next' 'Finish.')

On the left under the 'Titles Tab' you can name your chart under 'Chart Titles.'

'Category (X) Axis' lets you name the vertical axis. We usually leave it blank as a default meaning $ value, the price at which the stock has traded. (More about setting the axis values later.)

'Value (Y) Axis,' the horizontal axis, is usually left blank with Ticks as the default.

Switching to the Axis tab check only Category and Value and select 'automatic' on the radio button.

Now let's go to the 'Gridlines' tab and check 'Category (X) Axis,' Major Gridlines and 'Value (Y) Axis' Major Gridlines.

Last let's go to the 'Legend' tab and you can click on 'right' if you want the legend shown on the right of the chart. It's not essential.

We don't use the other two tabs.

Now click on finish – and lo and behold! You have a chart that looks something like this:

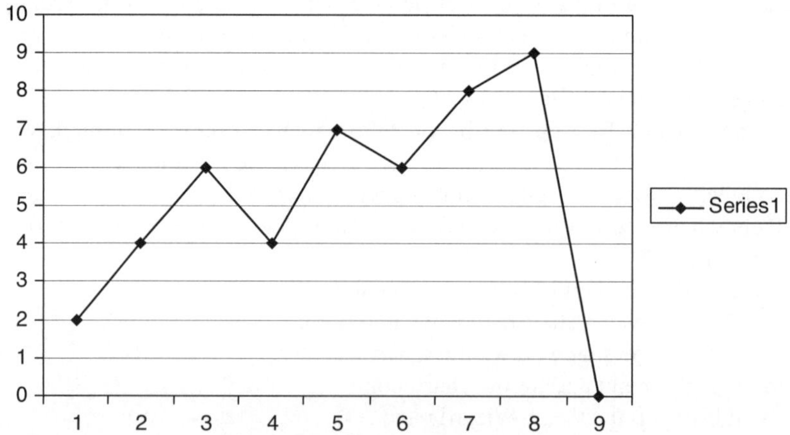

Figure 16.1 Excel chart

We had to change the default chart a bit to make it all in black and white.

The 'chart' area is the center rectangle and can be colored as you fancy. Our defaults are on the CD file 'CHARTS.'

The axes may be adjusted – right click in the vertical axis and this will bring up a small panel with the choices 'Format Axis' and 'Clear.' Left Click 'Format Axis.' This brings up the 'Format Axis' panel.

Go to the 'Patterns' tab. Select 'Custom.' This now gives you the facility to select line types, line colors, line width (all from the three drop down menus). The sample line on the bottom shows you the effect of your choices.

On the right three sub panels:

'Major Tick Mark Type' select 'outside'
'Minor Tick Mark Type' select 'none'
'Tick mark labels' select 'next to Axis'

Onwards to the 'Scale' tab:

'Value (Y) axis crosses at category number': Select 1
'Number of categories between tick-mark labels': Select 600
'Number of categories between tick marks': Select 600
[the 600 tick values we standardize on for most stocks]
'Value (Y) axis crosses between categories': Check (leave the rest unchecked.)

On the 'Font' tab:

'Font': Select 'Arial'
'Font Style': 'Bold'
'Effects': leave unchecked

If 'Auto Scale' is checked, uncheck it. This is important as Excel will proportionately change the type size when you change the size of the chart.

'Underline' = 'None'
'Color' = 'Automatic'
'Background' ='Automatic'
'SIZE' = select 12

The preview panel gives you some idea of what you are getting into.

(Only two tabs to go, don't lose heart . . .)

Go to the 'Number' tab. Scroll down to 'Number.' The default in the panel is two decimals – that's OK, leave as is for \$. If you are charting volumes reduce decimals to 0.

'Use thousand separator (,)': check.

Last tab here: you can leave all at default. Phew, done it. This needs a latte reward . . . (apologies to non coffee drinkers. . .)

Onwards: If you now left click on the area 'framing' the chart 9 small solid black squares appear, one on each corner and one each on the middle four containing lines. If you place your cursor in any one of these you can move the size of the chart.

Wider and narrower (black squares at the sides)
Taller and shorter (using the top or bottom black squares)
Corner squares to change the aspect ratio.

Click again anywhere in the 'frame' and hold down the mouse button – you can now drag the entire chart around the screen and position it where you prefer.

If you use a glide pad just touch down and hold down your finger as you move about the pad till you get the chart exactly where you want it then let go.

Double click anywhere in the plot area and this will pull up the color selection panel for you to select what color you want it to be.

You can also put a border around it (we don't usually).

Put your cursor in the plot area and right click. This brings up a minipanel with useful stuff:

Format plot area – as before
Chart type – options as before, can still change
Source data – vital – we will deal with it separately
Chart options – as before
Location – this gives the choice of 'embedding' in the current sheet or to set it up as a separate chart sheet. We always prefer the former.
3D View – not used much, which is a pity
Chart window – heads chart with current filename
Clear – hardly ever used.

To format a gridline put tip of pointer against it and right click.

If you want to add more plots to the chart right click in the chart area and select 'Source Data' and change tabs to 'series.'

This will bring up the 'Source Data' panel.

There is a neat trick for adding a new series to the chart:

Highlight the series already showing, Ctrl+C to copy it and then backspace to delete it.

Hit 'Add' and backspace the remaining contents in the bar, ={1}. Now Ctrl + V into the new blank bar what you had previously erased and change the column refs as required to ref the new series.

Trend Lines

A very useful feature in Excel charts is the ability to add 'trend lines' to a curve and to also be able to call up the 'fit' of the trend line for a number of common equations.

Place your pointer touching a curve and left click – a small label will appear with something like 'Series 1 Point 3' showing you exactly where you are on the curve. Now right click again to bring up a mini menu showing:

'Format Data Series'
'Chart Type'
'Source Data'
'Add Trendline'
'Clear'

Select 'Add Trendline' which will bring up the 'Add Trendline' window. We use the terms 'panel' and 'window' often interchangeably.

Here you see six equation choices, each of which describes a particular type of curve for which we may want to have a fit against the real live data.

The equations are: Linear, Polynomial, Logarithmic, Power, Exponential, and Moving Average. The moving average has a selection feature for how long you want the trend line to lookback. Its maximum is 255 periods (ticks).

If you have a linear-like line and want to know how far it is from a straight line tap on 'linear.'

$$\text{LINEAR} \quad Y = b + mX,$$

where b is the 'intercept' with the Y axis and m is the 'slope'.

$$\text{POLYNOMIAL} \quad Y = b + c\,x + c_2x^2 + c_3x^3 + \ldots + c_nx^n$$

where b and c are constants.

$$\text{LOGARITHMIC} \quad Y = c(\ln)x + b$$

where b and c are constants and ln is the natural logarithm function.

$$\text{EXPONENTIAL} \quad Y = ce^{bx}$$

where c and b are constants and e is the base of natural logarithms.

$$\text{POWER} \quad Y = sx^{b}$$

where s and b are constants.

Moving Average

This is the function we use almost constantly. It gives an immediate idea of how quickly the series we are looking at will smooth out and if there are any extremes showing up.

The maximum number which you can scroll up to in the lookback is 255.

We have one more rather important point on charts which needs careful attention: What do you do if you have an additional curve whose values lie outside the left hand X axis values?

Here is what you do to place it on the right hand axis.

Identify the Series number in the 'Name Box' (top left hand corner just under the Menu Bar) by scrolling using the up/down arrows. Or you can id the series by looking for it the same way but on the formula bar.

Once found go to Menu bar and select `'Format'|'Selected Data Series'` this brings up the 'Format Data Series' window. Select `'Axis'|'Secondary Axis'` this puts the series on the right hand axis. You may want to adjust the scales that Excel defaults to.

Please look up the File 'CHARTS' on the CD for some examples.

17
Our Metrics – Algometrics

Here is a description of our main metrics, all of which you will find as examples on the CD in the file we nicknamed 'ALGOMETRICS.'

Virtually all these metrics relate to the tick data with minor exceptions such as, for example, 'market capitalization' and 'shares outstanding.'

We use only a selection of these metrics as they may be required to illuminate specific questions. In all cases we can craft a series from the metrics which can take on any lookback period required from one day to 60 sessions, or more if required. We have not found it profitable to extend the lookback past the 60-session point at the present time and believe it more efficient to work on smaller session lookbacks and perhaps a larger variety of analyses.

Simple EOD series are useful for a rough and ready global understanding of the data. 20 EODs are a good start.

Fundamental analysis (where we would look at the company, its products in detail, the management, the balance sheet to evaluate the financial condition of the company, the profit and dividend history) is not being considered by us in our analyses as to all intents and purposes our trades happen in too short a time span for any fundamental consideration to have time to exhibit an effective influence.

However reading through a description of the company's operations and products will help cement the 'friendship' with the stock, which we have found to be quite valuable.

The objective of this chapter on Algometrics is to give you a range of tools with which to start developing a close understanding of the individual stocks. Learning to use the various metrics and seeing how they relate to the various ALPHA ALGOS will rapidly provide you with a 'feel' which most traders take a very long time to develop.

1. TRADE PRICE

One of the most important properties of a stock (obviously) is its current trade price tier location.

We have found that there are major differences in how stocks trade in the various price tiers. Anecdotally we have experienced considerably better results trading the higher priced stocks. A possible explanation may be that the lower price stocks attract less experienced traders while the higher price stocks are the domain of the Tier 1 players.

Price is one of the defining characteristics of a stock. It appears that the trading character of a stock is strongly modulated by its price. Besides the obvious proportionality issues ($1 return on a $100 stock is 100bp while that same dollar returned by a $20 stock is 500bp) the 'feel' of these two situations is totally different.

So one could assume that there are some differences in the trading characteristics.

Here are our Trade Price Tiers (please see 'ALGOMETRICS – Price Tab' examples on the CD):

1. Level 1 $10 to $30
2. >$30 to $50
3. >$50 to $80
4. >$80 to $120
5. >$120

2. %RANGE

We can calculate the %Range intraday of the stock over a number of windows and lookback periods. Our usual windows are 50T, 75T, 100T, 150T, 200T intraday (these can be sliding windows or 'boxcars') and EOD. See the examples on the CD.

This can usefully be extended over a lookback EOD of 5 trading sessions and onward to 60 sessions for some exceptional situations, preferably in boxcar fashion.

An average %Range/stdev is also of use with an EOD or when we use longer sequential intraday global lookbacks. Here we string say 100T sequential lookbacks over a global lookback period of 5000 ticks – this produces 50 data points which can then be plotted.

We can use %Range as another dimension of volatility so it can be considered as a proxy for a 'volatility' measure (one of many). All may to be plotted in Excel.

Formula for %Range:

$$\%\text{Range} = \frac{\text{MAX} - \text{MIN}}{(\text{MAX} + \text{MIN})/2} \times 100$$

where MAX and MIN are taken over the desired lookback.

Notice we use the average of MAX and MIN to 'normalize' and make the value comparable between stocks.

3. PATTERNS

The global EOD or multi-session tick series will nearly always contain useful pattern information. Use intraday series of 100T, 150T and 200T to see the series on each session. Please see the CD file 'ALGOMETRICS – Patterns' tab.

4. ABSOLUTE DEVIATION

We take sums +- FROM MEAN on a LB 100T and 150T.

5. SESSION VOLUME AS % OF SHARES OUTSTANDING

Session volume as % of Shares Outstanding may be more meaningful as a long lookback multi-session series to help assess if the activity in the stock is trending.

6. SHARES/TRANSACTION

INTRADAY on the usual selected lookbacks. Plotted as cumulative on 100T it may provide an indication of the Tier 1 involvement (you would see the shares/TXN ratio increase). Taken over long EOD lookbacks (20 plus) may on occasion prove useful.

7. RETURN

RET nT, n = 50, 75, 100, 150, 200, 250 and EOD. These should be plotted as line plots and also as histograms.

8. SIGMA RET

SIGMA RET on minimalist lookback n = 2 for EOD provides a comparative 'roughness' metric which can be used to compare the amount of 'jitter' in the stock price.

STDEV SIGMA EOD S. also nT = 100, 200, 300, 500T intraday.

STDEV RET 50T, 100T, 150T on EOD lookback.

SIGMA TRAVERSE EOD, 100T, 2T, 5T.

'CROSSING DENSITY' – as we look at the mean reversion characteristics of our tick series there are a couple of metrics which we find useful: The number of zero center line crossings per nT or nt. The amplitude of the reversion characteristic either as an average over nT or as a series to reflect increases or decreases in this value. As usual also EOD.

RET histograms in basis points 20T, 50T, 75T, 100T, 125T,150T, 200T. These can be carried out in small segments but we prefer to lookback at EOD. This provides some guidance on what the best 'holding time' should be and on what is possible for the specific stock being studied. You can also deduce what the best $stop ought to be.

LC ROUGHNESS INDEX

This is an absolute sum of 20T % returns with lookback to EOD.

18 Stock Personality Clusters

We shall continue the 'practical' part of this book with a brief exposition of our current views on what we call the 'personality' of stocks and a brief overview of some principles of clustering.

Partly due to its wide application span and multiplicity of measures of proximity 'clustering' is not that easily defined. We define (restricted to our usage) cluster analysis as the art and science in data analysis in which stock symbols can be assembled into groups of varying 'tightness,' proximity, or closeness of similarity on the properties, metrics or traits of your choice.

To put it in slightly more general terms: Classification of objects ('objects' would include mathematical series, mathematical properties, time properties, magnitudes, industry segmentation, even thought patterns, anything you could define) into different sets according to their degree of proximity (closeness or its inverse) on a selected metric. When we head for multiple metrics things can get quite complex when dealing with stocks, so normally we work with one metric at a time.

Clustering may also be used for understanding the natural structure of the data. It may be used as a summarization tool. It may be used to define and provide useful subsets. It may be used to prove the 'belongingness' of a controversial item to a set.

Our prime conjecture is that stocks of close proximity are likely to exhibit similar trading characteristics. The proximity dimension of the cluster members is considered to provide a good indication of their property similarities.

This conjecture, that sets of similar objects will have similar behaviors, at least to some extent, has no rigorous proof but appears to behave up to expectations reasonably well – in some instances the parallelism of behavior is quite remarkable. So that if we manage to solve the problem of price movement trajectory for one member of the set we may have also come closer to solving it for all similar members of that cluster set.

Though the concept of 'clustering' is extremely powerful, the process suffers from a large dose of arbitrariness as any sample data set can be partitioned in a variety of ways with little reason for the preference of one partition solution to another. Thus we will still need further elaboration of our model to choose appropriate metrics on which to prioritize.

From a mathematician's point of view clustering is an 'ill-posed' problem as there are many solutions with no reason to prefer one to another. Clustering techniques can be strongly influenced by both the strategy adopted as well as the biases and preferences of the operator. Skill and experience make all the difference to obtaining successful results.

Stock 'personality' clustering is the subject of our ongoing research work. We are carrying out work primarily on the S&P 500 and some selected ADRs.

We usually start to analyze a stock's 'personality' using a limited set of metrics we have developed and standardized. A good start is the '%RANGE' metric. Here is a 'toy example' looking at one day only:

	STOCK	METRIC
		%RANGE
1	AAPL	0.056
2	ADBE	0.019
3	ADSK	0.021
4	AKAM	0.027
5	GOOG	0.013
6	ALTR	0.018
7	YUM	0.018
8	AMAT	0.027
9	AMD	0.061
10	AMGN	0.020

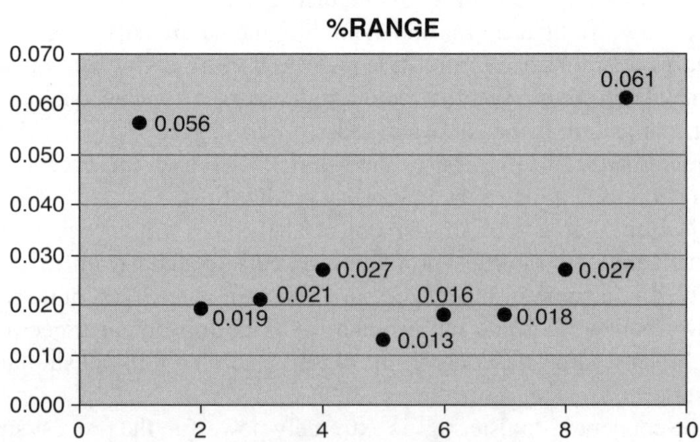

Figure 18.1 Example of '%RANGE' metric

You may have noticed the 'stripe' around 0.020. This is a cluster we trade often. Notice also the distance from this cluster to GOOG at 0.056 and AMD at 0.016 both to be 'handled with care.' They trade differently, sometimes much faster in $ per second.

Before we dive any further into the metrics here is a brief discussion of cluster analysis. First, a warning – as previously mentioned cluster definition is an 'ill-defined' or 'ill-posed' problem as the clusters very much depend on the context, and primarily on the ability, bias and goals of the analyst.

We always say, half in fun, that what we call 'eyeballability' is more important than the math. This is also very much the thrust and opinion of Professor Mandelbrot whose words I take the liberty to paraphrase – 'you have some wonderful tools developed through millennia of evolution – your eyes – use them.'

The inputs for the data to be cluster analyzed are taken from a backtest historical tick dataset. For a reasonably comprehensive analysis we suggest this should be comprised of between 20 and 60 sessions of the NASDAQ and NYSE Exchanges using your selected Watchlist ticker symbols and tick data as recent as you can get.

Be careful not to underestimate the task: 100 stocks times 60 sessions gives you 6000 files so be prepared for some late nights. You may find some help in our 'METRICS' file on the CD which may help cut down the work load a little.

The underlying idea is that different stocks appeal to different traders – different people segments like to own particular stocks without any great logic underpinning the preference. The follow-on logic argument is that these stocks will have attributes, perhaps reflecting the owners' or traders' psychological makeup and their preferences which can be arranged into coherent clusters.

And, taking the logical skein further it could be conjectured that the stocks in a tight cluster might respond to the same algos and that they are also likely to move together, more or less, or at least exhibit some properties which we can use in the design of our algos.

Thus by analyzing stock properties and their behavior clustered on various metrics we will hope to develop more or less homogenous cohorts of stocks which will respond similarly and where selected algos will have a higher probability of producing excess returns.

This very clustering process also provides thought for new algorithmic strategies which explore and possibly make use of some particular facet or feature which a particular stock exhibits.

Once we have chosen a property or metric on which to cluster a Watchlist of ticker symbols we have a choice of a large number of clustering methods. We usually choose what is termed the Euclidian distance method which has the nice property of allowing us to use Excel scatter diagrams besides the actual math.

The formula for computing the Euclidian distance D, between two values of a stock variable metric is:

$$D_{ij} = \surd(X_i - X_j)$$

where D is the 'distance' between the two stocks and their metrics.

The smaller the value of 'D' the closer the 'proximity' of the two stocks.

Just a simple scatter plot in Excel is usually quite adequate and a lot simpler. Please see examples on the CD in the 'CLUSTERING' file for an indication.

Some concluding driving directions:

You want to partition your stocks so that data that are in the same cluster are as similar as possible.

You want to achieve maximum heterogeneity between clusters – symbols that belong to different clusters should be as 'different' as possible.

As shown in the example above, using only one easily calculated metric is quite an adequate tool, even when used very 'lightly' without the rigorous math.

Please be aware that a really crisp and clean cluster assignment is not always possible and we have to make do with the best that is available. Much of the clustering will tend to be related to sector membership.

Just in case this methodology has wetted your mathematical interests Wishart (1969a) has developed a method seeking disjoint density surfaces (regions) in the sample distribution which he named 'Mode Analysis.' This is really just a much more powerful but complex clustering method.

The search through the data is made by a sphere of radius 'R' surrounding each data point and then counting the number of points which fall in the abovementioned sphere. Individual data points are then labeled as 'dense' or 'non-dense' depending on whether their spheres contain more or less points than a linkage parameter, 'K' which depends on the magnitude of the dataset. (Wishart (1978) suggests some possible values for 'K.')

If you really get stuck in the 'clustering space' there is no shame in resorting to clustering software to help do the job. We have used the software package 'Clustan.' David Wishart created it and has always been most helpful: www.Clustan.com. Department of Management, University of St Andrews.

We have not used these but they are highly recommended: www.ncss.com and www.visumap.com.

19
Selecting a Cohort of Trading Stocks

This chapter lists our first preferences in stock sectors we have experience with. The full Watchlist of stocks, most of which we have traded, together with some basic outline information and company descriptions you will find in Appendix B and Appendix C.

We have our modified industry sector list in Appendix A.

Our definition of a cohort: Symbols which we have found to have similar trading characteristics when they are defined by our metrics.

Here is a first, cursive narrative routine to get you started. The file 'COHORTS' on the CD lays out the routine in Excel.

The Price Tier metric is our first port of call. If there is sufficient capital in the account we have found that we have had more success trading the higher priced stocks. We speculate that the higher priced stocks have a totally different ownership profile to that of the lower priced stocks.

Trade 500 share lots as a stop gap if there is not enough cash in your trading account to comfortably cover at least two 1000 share trades at the same time. A reasonable start is to pick a stock in the \$75 to \$100 list.

Price is only one of many dimensions on which we can 'fingerprint' a stock and connect with its trading characteristics.

The volume level metric is the second stop on the tour. We are using volume here as a proxy for liquidity. Do not trade any stock which has a three-session average share volume of less than 1 million shares per day or has one of the three sessions at less than 750 000 shares.

The %RANGE metric requires careful handling. This must be looked at in context of other stocks in the sector – it also needs a series approach to see how the average daily 200T %Range is moving. An EOD up trend over the last 5 days could mean that this will provide more trading opportunities.

Intraday the %Range 200T is a good indicator of 'tradability,' other factors remaining neutral. Ideally it should be over 0.002 at least half the time.

Do not trade stocks which have an EOD %Range of less than 0.002.

The traverse metric is a total of squared returns which shows the amount of 'movement' in the stock price EOD. It may be worked up to define a 'roughness' characteristic. The classic version used in this metric is $n = 2$.

Another version of this metric measures the price motion, or absolute traverse. We measure this over nT where $n = 100$ and 200 ticks which is then transformed into series. We also take the same measure at EOD.

Over time one inevitably tends to collect favorite trading stocks. This is totally individual – no two traders will have matching preferences. These preferences evolve from your trading experience.

We have drifted toward the oil and energy sector (e.g. RIG, OIH, FWLT, NOV) with the pharmaceuticals (PFE, AMGN, BIIB, CELG) a close second. Our global favorite stocks are CME, GOOG, and SHLD.

20 Stock Profiling

Each stock that you trade should get a dedicated Excel file to archive all activity, results and the trade data. Over time this will become a valuable database resource. We have included a sample of the Excel sheet we use on the CD, in the file 'STOCK PROFILING.'

Pay special attention to which algos seem to produce better results for the main optimizing criterion, bp/sec. (Trade return over trade duration.)

If the OMS you choose has the facility to copy the transaction log to your hard disk, do so at the end of each session.

We target the basis point return at a minimum net of 25 bp, which means that you have to calculate the winning less losing trades and also subtract the brokerage commissions to arrive at a true net figure.

Note the average brokerage commission breaks at a round trip on up to 2500 shares. After 2500 shares you pay a fractional price per share extra. Make sure to check your broker's tariff.

Please remember the Session Volume 'smile' characteristic of most trading volume in most sessions. This shape is due to higher volumes at the Open and at the Close. At the Open it is driven by 'price discovery' as the market seeks to assimilate the overnight news. At the Close it is driven by traders having to complete or close out positions.

This does not mean that our ALPHA ALGOS are designed not to trade at the 'outskirts.' We are basically time agnostic as our time slices tend to be small.

Many practitioners, however, prefer for the market to 'shake itself out' in the morning before they put on a trade, as things can get pretty hectic at the Open. 15 minutes is frequently quoted before putting on the first trade of the day, with some practitioners holding off trading even longer until the so-called 'price discovery' process has stabilized. This may take an hour or more, till about 11:30 am on some days. On the other hand the large swings happen often at the Open, so if you are on the right side of them, they can be most profitable.

The algos are designed to take on the turbulence (pay close attention to the speed of reaction of the stop loss) as the swings may be very rapid indeed at the open. If we have tested them thoroughly on the target stock we can simply let our algos tell us what to do.

21 Stylistic Properties of Equity Markets

We will briefly describe some facets of the markets, often called 'stylized empirical facts.' The explanatory hypotheses of what we believe causes these stylistic properties are left for another day.

A 'volatility turbulent' day is usually followed by a second volatility turbulent day. Not quite symmetrically – a 'tranquil' session is often followed by one or more tranquil sessions.

Volatility is usually higher following a down market day. High volatility events tend to cluster in moderate (days) timescales.

Among a number of other theories it is thought that volatility 'switches' up or down between a small number (2 or perhaps 3) distinct states or 'regimes.' The regime model proposes that volatility switches between 'high' and 'low' states but gives us no hint as to how long it stays in each state or what conditions create the regime switch. This phenomenon is usually combined with 'volatility clustering.'

Generally, over a test sample of 100 selected stocks the high frequency (at intervals of one minute or less) returns are not Gaussian, that is, not normally distributed but Gaussianity seems to reappear as we increase the timescale over which the returns are calculated.

The distribution of returns has monotonically increasingly fat tails as the sampling frequency is increased. We have found that this feature also varies in magnitude among individual stocks.

There is a varying amount of autocorrelation of returns from stock to stock which also seems to be specific to individual stocks. The autocorrelation effect seems to wane then disappear after 180 to 300 seconds. When the microstructure of the market is not engaged autocorrelation of returns is insignificant at sampling rates over 600 seconds.

This autocorrelation at high frequency sampling is an important part of designing our ALPHA algorithmic strategies as most of our triggering activity relies on autocorrelation as part of the pattern recognition process.

Various scaling laws can be used to describe absolute and mean squared returns as functions of their time intervals. It would appear that there is a proportionality relating to a power law of the sampling frequency.

There is some considerable asymmetry in the movement of prices with the drawdowns being significantly larger and faster than the upward movements.

There is considerable variation of price trajectory smoothness both as stock-specific features and also as variations in the performance of individual stocks over time in relation to the market as a whole.

We have found a certain amount of predictive power in the bellwether stocks of a sector. It appears that traders take a finite amount of time to take in the movement of these stocks and only then decide to look at the related stocks in the sector.

This may also be a feature of the information diffusion process. We might be tempted to assign some form of 'time constant' to it but so far we have too much variability in the data pointing to there being another variable (undefined so far) hiding in the mix. There may also be more than one major variable that influences things which we have not yet been able to track down.

Professor Vandewalle has shown some interesting topological relationships between stocks and how their returns are correlated. We have extended and applied some of these concepts to one of our algos (General Pawn). The related website www.market-topology.com shows much work on correlation of stocks and has some very good topological displays of stock relationships.

We observe decreasing kurtosis (which is the height of the normal distribution: the tall ones are called leptokurtic, and the flat ones are called platykurtic) in return distributions as the sampling frequency is decreased and the interval between the sample points is increased. At intervals of over five trading sessions we found the kurtosis approaching the Gaussian values around three, whereas at high frequency intervals (60 seconds or less) the distribution was distinctly leptokurtic.

The Law of Large Numbers: as more than 30 or so samples are taken from a non Gaussian distribution the distribution of the samples is Gaussian! Go figure . . .

Let us have a look at how the size of returns relates to the frequency at which we measure it. Please see the 'RETURNS' file on the CD for details, some of which are shown in Figures 21.1 to 21.3, below.

The data column was selected at random from our tick database and contains 5714 rows, which we shall use rounded down to 5000.

This has two inbuilt timing parameters:

1. The tick window interval on which the return is taken.
2. The total size of the lookback over which it is calculated.

It is important to note that the returns exhibit 'mean reversion.'

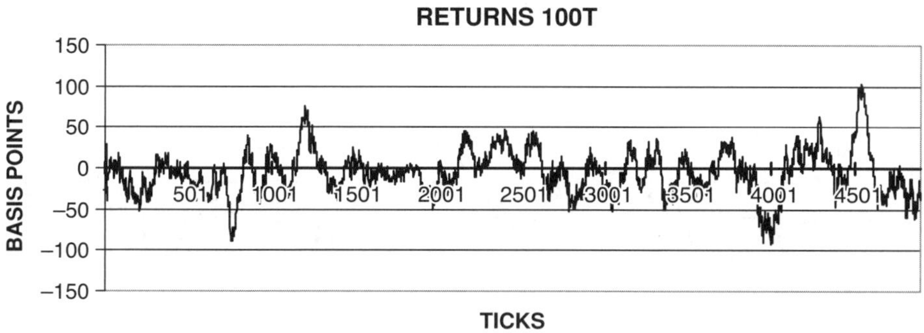

Figure 21.1

From a trading viewpoint we can see what the holding periods will look like at a return level.

We can also get a view of the absolute maximum number of highest returns at a chosen holding time.

The following examples illustrate the influence of the tick interval length (trade 'duration') on return magnitude. Besides these examples please also study the examples on the CD file 'RETURNS.' Tick return analysis is one of our more powerful analytic tools. Extracting the maximum information from a chart is however more art than science. Being very careful and super-observant helps. Make sure you check that the 'y' axes of comparable charts are the same.

There is much to look out for. Notice the bottom two charts in the 'Return Charts.' Just by changing the aspect ratio we get a different impression.

In Figure 21.1 we use a return interval of 100T over a total lookback of 5000T.

In Figure 21.2 everything in the previous chart format is constant so that we can evaluate the comparative changes due to simply extending the return period to 150T.

Figure 21.2

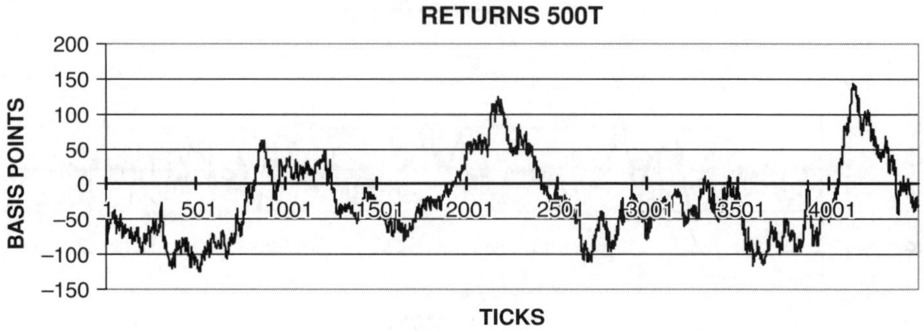

Figure 21.3

Notice the rather deeper excursions obtained by changing the return period to 150T. The detailed shape also changes rather subtly.

Figure 21.3 uses a return interval of 500T over a total lookback of 5000T. Thus we have a return series which has a much larger window over the same lookback.

Without much analysis we see immediately that the differences in the return pattern are quite drastic notwithstanding the sliding window smooth. This return length is longer than we might usually trade and shows an exaggerated view. We have included it really only to emphasize the importance of this analytic tool

As you can see from the above rough examples returns are very much at the mercy of the 'holding time' in this microeconomic area.

This now opens up the Pandora's box of trying to establish an optimum holding time to maximize the return on the trade. Both volatility and returns are precise, quantitative concepts.

One could think of volatility as firmly entrenched in the lookback, while returns are very much a function of the tick length.

Obviously return versus holding time is not a linear function as we are looking at convolutions of various drivers of various frequencies and amplitudes, all of which also vary in their own right. Complexity at its best.

It is quite surprising in some ways that many practitioners often judge the 'riskiness' of a stock primarily from the perception of the irregularity of the price trajectory of the stock price and its 'jaggedness' to some extent as a proxy for their conception of 'return' and 'volatility.' What they are really seeing is the convolution of all the drivers of the price dimension.

22 Volatility

The word 'volatility' is probably the most used word on the Street and in common by people discussing finance. Few of the users of the word have even the slightest idea of how complicated and varied the concept really is. The best exposition of the topic we have read is by Professor Ser-Huang Poon (2005) (please see the Bibliography for details).

We shall skim this difficult topic, restricting our aim to its use in our stock selection and try for some slightly deeper understanding of how the concept impacts the strategies of algorithmic trading.

At the most fundamental level volatility can be defined as the price change of a stock, the fluctuation, per unit time.

We express it as basis points per second or basis points per nT ticks where n is usually = 100.

An important property of volatility is that it is 'mean reverting.'

'Volatility' and the concept of 'returns' are closely linked.

Volatility can be expressed on many timescales, from the smallest to daily, monthly, even annually depending on the context in which it is being used.

Volatility is present to some extent in all market conditions. Whether the market is trending up or down or just moving 'sideways' there are always 'ripples' of varying size overlaid on the base.

As a rough rule of thumb markets trend up 20% of the time, and down 20% of the time, with the rest 'sideways,' but, as a rule a level of volatility exists in all market conditions, to a greater or lesser extent.

The estimation of volatility is very much dependent on what your requirements are. The best explanation for our purposes is that it varies with both the lookback period over which we want to calculate the volatility and the sampling frequency we decide to use.

We often use the finest sampling resolution and a number of sequential lookback periods (these can also be of various lengths depending on the market activity of the

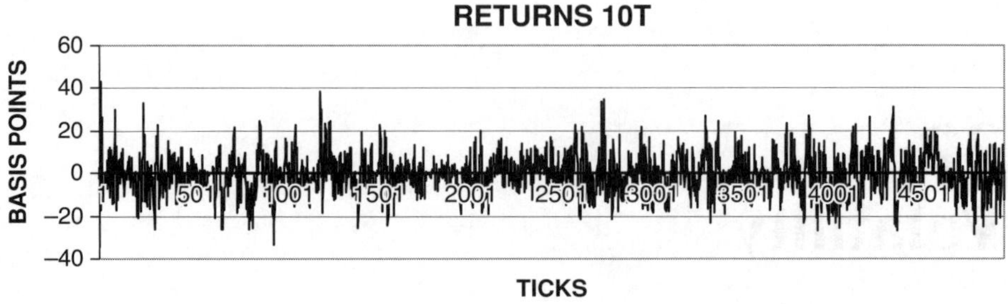

Figure 22.1

stock we are analyzing) to observe the variational behavior of volatility over time, usually over three to ten trading sessions.

We also use EOD time series methods to track the volatility of a stock to assess if it is increasing or decreasing or indeed if the volatility varies substantially over time, a kind of volatility of the volatility . . .

With tick data commonly available, the term 'realized' volatility is now in common use to refer to estimates using intraday squared returns at 'short' sampling intervals of say five minutes. In our methods we use shorter intervals, mostly expressed in tick values, equivalent more or less to 60 seconds or less.

An important feature is that the variations appear often 'bunched' together. These short periods of increased amplitude are usually referred to as 'volatility clustering.' Figure 22.1 uses 10T interval returns to illustrate the clustering behavior of the series. The darker areas are the 'clusters.'

As previously mentioned an unexplained property is that a high volatility clustering day is often followed by another high clustering day, whereas a quiet period is usually followed by another relatively tranquil period.

Volatility is often calculated and expressed as the standard deviation of returns.

$$\sigma = \sqrt{1/(P-1) * \Sigma(r_P - \mu)^2}$$

where

P stands for number of periods,

r is the period return and

μ is the average return over the lookback period.

Thus: sum the difference of each return from the mean, then square it, then divide it by the number of returns less one and take the square root.

In addition to actual standard deviation of prices, which is also often used, and the return volatility described above we also use a 'high – low' 'range-based' (sometimes called the 'extreme value' method) where we have a number of preferred very short lookbacks (often in the range of 200 to 500 ticks) in contrast to normal usage which is usually daily.

Of all the various volatility estimators we usually prefer the %Range metric intraday averaged over 200T because it allows us to compare metrics between various stocks.

We often generate a time series from these to estimate the change in volatility of the stock over time. This is usually calculated over anything from one session intraday with lookbacks of say 100T all the way up to 20 trading sessions.

Please have a look at the CD file 'VOLATILITY' for examples and comparisons.

23
Returns – Theory

Here is a small sample of what we have discovered so far when looking at returns with various windows.

When we examine the behavior of the absolute size of returns they are a function of the frequency at which we measure them. The lookback length (its limits range from 'boxcar' to EOD) on a stock with 30 000 transactions could be set at 1000 and return at 200T.

The definitive formulation of the generating function is still a matter of some heated debate and controversy. Not really too surprising when we take into consideration the enormous complexity of the task.

The CD file 'RETURNS' shows examples of the effect of return length. Take your time studying the charts as they are the route to a much deeper understanding of the markets.

From the analyses we can make various guesses as to how long we may need to hold a trade and what the 'perfect' return would be and generally get some idea of what the probabilities are.

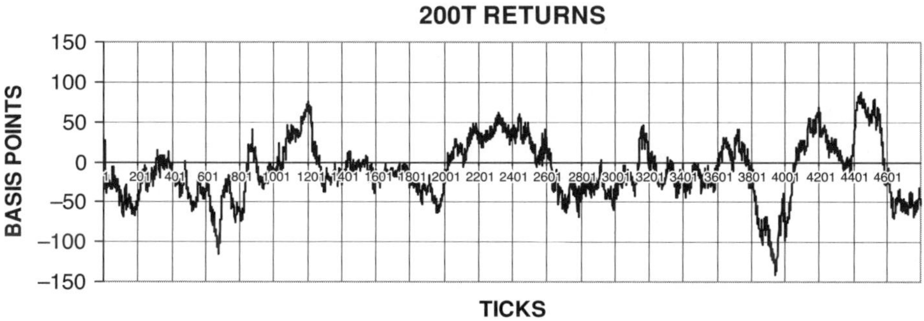

Figure 23.1

An 'optimum' time cannot currently be derived *a priori* and must be guessed at from recent data. The optimum holding time fluctuates in synch with the current volatility of the individual stock. Stocks in the same 'cluster' appear to have similar 'optimum' holding times.

The vertical gridlines are 200T spaced. This gives us a rough idea what a particular holding time would result in.

24
Benchmarks and Performance Measures

Performance measures are useful in comparing performance between different stocks and algos:

VIX
Beta
Alpha
Sharpe ratio
bp/sec

The VIX is more of an indicator than performance measure. Beta is a measure of volatility against the S&P500. Alpha is a measure of outperformance of the index. The Sharpe ratio measures risk adjusted performance. Our favorite is bp/sec.

THE VIX

The VIX is an Exchange traded indicator of the implied volatility in the trading of Standard & Poor's 500 options. High values of the VIX happen to coincide with periods of market uncertainty, fear of loss, and general nervous sentiment.

A low VIX reading usually may portend a market rise, low volatility and general calmness among market participants.

The VIX is best interpreted at the extremes of high or low reading where it tends to be an accurate predictor of events to follow.

BETA

Beta, symbol Greek B, β is the relative volatility of a stock against the volatility of a benchmark, usually taken as the S&P 500 index to represent the entire market. A $\beta = 1$ means that the stock has the same volatility as the S&P 500 index.

A beta of greater than 1 means that the stock fluctuates more than the benchmark. Similarly a beta smaller than 1 means it fluctuates less than the benchmark. Thus a beta of 1 means that the stock moves in parallel with the market.

For example if the stock has a beta of 2.2 and the index moves up 2% we would expect the stock to move up 4.4%.

A stock with a beta of 1.24 would be 24% more volatile than the benchmark.

Cumulative weighted betas are used in portfolio analysis with the aim to diversify out (reduce) the stock-specific volatility/risk with the aim to get as close to a figure of 1 as possible.

Betas from 0.5 to 4 relative to the S&P 500 can be seen depending on stock and sector. The calculation to obtain beta is a simple linear regression of daily returns of the stock regressed on the S&P 500. It is usual to use a rather long lookback period, say one year, though we have used it with short lookback periods to create a 'beta series' to analyze the trends for specific stocks.

The regression analysis may be used to find the relation between any two series, thus we could look to see how one stock's volatility varies against that of another stock.

ALPHA

Alpha, the first letter of the Greek alphabet α, is a measure of volatility/risk discounted performance. It is the stock return when the benchmark return (usually the S&P 500) is 0.

Following on the logic it is obvious that, if a stock's beta is 2, and the benchmark return is 2%, it would be expected to gain 4%. If the stock gained 5%, it would have a positive Alpha of 1%.

The term ALPHA we have used in a generic fashion to indicate that our algos are designed to achieve high profitability.

The volatility to risk relation is in our opinion rather strained as there are so many factors to consider and much of the calculations are lookback-sensitive for both stock and benchmark.

An accurate and precise quantification of risk is in our view quite complex and possibly not even achievable.

Alpha can be easily calculated from the formula:

$$\alpha = \mathrm{RET}_{\mathrm{stock}} - \beta * \mathrm{RET}_{\mathrm{Benchmark}}$$

This is best explained by example:

A stock with a beta value of 2 provides a return = 12% while the benchmark index gives a return of 5% in the time span. A 2 value of beta multiplies the benchmark to give 10%, subtracting from the actual stock return of 12% gives us an Alpha value of 2%.

The Sharpe Ratio

This ratio was developed by Nobel Prize winner William Sharpe who shared the 1990 prize with Harry Markovitz. The intention of the Sharpe Ratio is to measure risk-adjusted performance.

It is calculated by subtracting the risk-free rate of return from the actual returns and dividing by the standard deviation of the actual returns.

For our use in trading ALPHA ALGOS we have simplified the formula to just the average of the return windows in the lookback period divided by the standard deviation of their sum. (The risk-free return is eliminated due to the short return periods.)

The idea of the Sharpe ratio is that it should indicate if the result of a good trading strategy is the result of smarts or the taking of excessive risk. Unfortunately it is most sensitive to the frequency of trades – please see Irene Aldridge's article starting on page 46 of this book.

The higher the value of the Sharpe ratio the 'smarter' was the trading strategy. This works only when comparing like for like return periods which is mainly of interest for long hold investors.

The Basis Points per Second and Basis Points per Trade

These are metrics we have used in order to have some sort of comparison measurements.

Bp/sec is an absolute cumulative metric which evaluates the profitability performance of a specific ALPHA ALGO in relation to time.

The same but measured and plotted over a complete trade gives a slightly different view, for example the algo may have run 150 seconds with an average profitability of 1 basis point per second. However within this period there may have been a point where it hit 30 basis points per second while others had 0 basis points per second.

The Excel templates can be set up to provide a real-time running value of the cumulative bp return for each trade.

25
Our Trading Algorithms Described – The ALPHA ALGO Strategies

1. ALPHA-1 (DIFF)

Motivation: Use convolution of low-pass and high-pass filters on price tick series to obtain a trigger signal line which can be parameterized to produce Buy and Sell triggers.

During the course of development we had rollercoaster-type swings in our returns. Due to our adaptive stop loss strategy we were spared any meaningful drawdowns. This LC stop saves things when the algo is not producing the desired triggers.

On the other hand we have had sessions with win/loss ratios of 8 out of 10 for several days in a sequence. Returns vary considerably – we target an average net of 25 basis points. The rare 300 basis points is a gift.

The most concerning times are when nothing is happening on the monitors – no triggers. You start looking at the connectivity, checking the UPS, checking the Time and Sales and generally getting nervous. This market/algo performance is a characteristic we have encountered frequently enough to have become used to it as a feature of the system.

The ALPHA-1 is a fairly consistent algo but requires quite a lot of time studying how various parameter settings affect results for particular stocks. Parameters are best reset daily. It is deceptively simple in its basic form:

1. Choose a lookback for an SMA based on the TXN number (we usually take a five-day average) and divide this average by an integer 4 to 7. On some stocks things look right quite quickly, on others you have to do lots of iterations to make

the SMA line look anywhere near the example on the CD file 'ALPHA-1' (white line on chart).

2. Multiply the value you obtained in the prior step by an integer, usually 6 for most stocks. This gives you the LMA (red line on chart) lookback value. Again it requires empirical testing for each individual stock to get the best results. The main choice criterion is that the LMA plot should be 'relatively' 'smooth.'

3. Subtract the value of the LMA from the SMA to generate another series we call the 'trigger line' (green line on chart)

The green 'trigger line' has its values on the right hand axis of your chart.

4. Using a longer lookback chart, say five days, though even three days will do on most occasions, establish a straight line value where the total valleys penetrating it are between 5 and 8.

None of these parameter producing numbers are cast in stone. They are not to be considered magic numbers, just fairly average heuristics, and you must exercise lots of trial and error to make the final selection for each stock. And keep reviewing it frequently (daily?).

5. Press F9 in Excel to ensure the spreadsheet is recalculated.

6. Now we plotted the SMA, white, LMA, red, trigger line, green as an 'embedded' chart in Excel in tab 'ALPHA-1 CHART.' Note that the trade price line is dark blue on the chart with its values in US dollars on the left hand axis. This is a classic template chart and worth detailed and thorough study. Take your time.

We will walk you through the construction of this algo in detail again in text in the next section 'The ALPHA-1 Algo Expressed in Excel Function Language' and on the CD file 'BUILDING ALPHA-1.' We will provide you with a test chart and data so you can generate examples of how the parameter settings affect the trigger line and demonstrate the sensitivity of the lookback dimension settings.

One of the unusual features is that the trigger line seems to exhibit very good fits to various parabolas. The closer the fit the better is the ensuing trade in most cases (good example of a heuristic). However, please remember that in real time you only get to see half of the parabola down to its vertex for downpeaks. The other half is still in the future.

The other feature of the trigger line is that there appears to be a strong relationship to the depth of the valley with the ensuing 'goodness of trade' of the Buy trigger.

Notice the red horizontal 'zero' line. This can be used as a condition for generating Sell triggers – the up-peak must be above this line.

Keep hardcopy records of parameters set: SMA, LMA, Buy level and Sell level and archive in your 'STOCK PROFILING' file.

Even though the results achieved vary considerably this is a good illustration of how algorithms verging on marriage with heuristics provide a trading edge.

It is possible that research which looks at the system from different perspectives could provide some harder guidelines for setting parameters, thus improving the efficiency of the algo.

Study of relationships with the %Range metric could be of value.

1.a THE ALPHA-1 ALGO EXPRESSED IN EXCEL FUNCTION LANGUAGE

Here is a first: us describing our ALPHA-1 algo source code, in print!

We will describe the construction ALPHA-1 component by component. Please open the CD file 'BUILDING ALPHA-1.'

Suppose that we have decided on the lengths of the SMA and LMA to be 400 and 1600 ticks, in white and red, respectively. We write them with five zeros to flag that they are 'live' parameters which we can change to see the effect on the trigger line in Col I.

We write the lookback value SMA into Column G, Row 1 and the LMA into Column H, Row 1. We will now add the required function code into the top row of each of the columns which will be calculating the moving averages from the incoming data in Column F.

The actual code going into Column G, Row 400 (which is where the SMA starts) is a bit complex unless you are quite familiar with the Excel function language. What it does is pick up the value for the SMA from Column G Row 1 and inserts it in the calculation of the average from there onward.

The general template formula shown for your information as text in Column G. Row 4 is shown below:

```
=IF((ROW()>g$1, AVERAGE(OFFSET($F4, -(G$1-1),0,G$1)), "")
```

This is what it does: If you write the lookback value, or SMA length into Row 1 of Column G when the data columns start scrolling down the spreadsheet, as soon as the data reaches the row given in the lookback value Excel starts calculating the SMA and writing it in the successive rows of the column.

So the actual code in Col G Row 401 is:

```
=IF(ROW()>g$1, AVERAGE(OFFSET($F401,-(G$1-1), 0G$1)), "")
```

which then uses relative addressing as it scrolls down and calculates the moving average down the column.

Parallel procedure is followed in Column H with the LMA but obviously with a different value for the lookback (1600 in this case). So the LMA will not start calculating till the data has hit the 1600th row.

So far so good. Hope you are still with us. It's a lot simpler to do than to describe in words of one syllable or less.

The very next job is to calculate the 'trigger' line of the algo. This is easy – just subtract the LMA from the SMA.

In Excel function language you write into the cell Column I, Row 1600 the following: '=G 1601 – H 1601' and then copy this formula down (don't worry, Excel has a neat way of doing this – place the cursor to the bottom right of the cell and a little cross will appear, just click and the formula copies down, automatically adjusting the relative addresses in each row, as far as its left adjacent column has data in it).

Now please go to tab 'CHART 1.' The usual color coding is used with some trigger point conditions from an ALPHA-2. The ALPHA-1 trigger line is shown in bold pale green.

In order to start comparing triggers we have also added in the ALPHA-2 EMA treble line in light green and shown where this line would fire Buy (white filled blue circle, dark blue 'legs') signals and when it would fire Sell (pink filled red circle, with red vertical 'legs') signals.

Compare these carefully. Notice that the various convolutions involved provide sometimes very similar results. The ALPHA-1 trigger often fires a bit later than the ALPHA-2.

Please turn to tab 'CHART 2.' Here we show two EMA lines. The aqua color is 200T and hugs the price line very closely. The pink/violet is a 600T EMA and is much smoother but does not follow the price line that closely and gives us many fewer peaks.

For the sake of experiment we have included a DIFF line (the 200T EMA less the 600T EMA) in dark brown. Notice that the DIFF peaks are much sharper.

Going back to tab ALPHA-1 we have added Cols AA, AB, AC which are another instance of SMA, LMA, DIFF which are a 'spare' you might want to experiment with by changing the values, currently set at 200T and 1200T, to see what difference variations make. Tab 'Chart 3' is set up with a test chart which will reflect any changes you make.

Be prepared to spend quite a bit of time and concentration on assimilating these patterns, how the trigger lines show trading opportunities, how the slight difference in the math makes a large difference in what you see on the charts.

There is no substitute for taking your time on this chapter. It will provide a solid foundation for the rest of your work.

2. ALPHA-2 (EMA PLUS) V1 AND V2

Motivation: Use EMAs of various lengths to smooth the price line and use EMA stacked diffs to provide trigger signals.

This algo has two versions.

V 1 uses only one EMA to trigger Buys on valleys and Sells are signaled when the indicator line has travelled upward over the red zero line and provides a up-peak. Start by trying out a 300 EMA and proceed from there.

V 2 uses three EMAs with the third EMA being run on the difference of the first two. Please open the CD file 'ALGO – 2 V2.'

The stock price is in Col A.

Col G, Row 2 is the lookback value, 200T, for the EMA calculation in that column. The bold value 0.009950 is the EMA constant (often confusingly called 'alpha,' α). It is calculated as

$$\alpha = (2/(\mathrm{n} + 1))$$

where n is the lookback, 200T in this case.

$$\mathrm{EMA} = \mathrm{EMA}_{t-1} + ((2/(\mathrm{n} + 1)) * (\mathrm{S}_t - \mathrm{EMA}_{t-1}))$$

is the equation for the EMA calculation.

Similarly Cols I and J with 300T and 500T lookbacks.

Col L is the difference between Col I and Col J.

Col M is the Trigger line and is produced by running a 200T EMA on the Signal Line in Col M.

'BUY TRIGGER CONDITIONS FOR TRUE,' the resulting 'TRADE PRICE' and the 'SELL CONDITIONS FOR TRUE' with the resulting 'TRADE PRICE' are calculated in Cols N, O, P and Q respectively.

It is useful to remember that if you want Excel to show you a formula and where this formula is getting its values from you just press 'F2' ('Esc' to cancel).

Another handy shortcut (which we have already covered but is so useful that it bears repeating) if you want to remind yourself of the structure of a formula is to precede it with an apostrophe. It is then turned into text and is ignored in calculations. You can turn it back into a formula by deleting the apostrophe.

We will not 'parse' the code in general except to show you how we identify up-peaks and valleys. Col M is the line where we want to tag the peaks. The following code finds the valleys and thus the 'BUY TRIGGER CONDITIONS FOR TRUE.'

```
=IF(AND((M27<MIN(M24:M26)),(M27<MIN(M28:M30)),M27<−0.04), 1, "")
```

This code will find trajectory reversals (valleys) with a lag of only three ticks.

If true the formula will write a '1,' if not, then it does nothing. This '1' is the 'BUY' signal. The next Column contains a continuous check for this '1.' If TRUE then it fetches the current price for which the trade should be made and writes it to the parallel cell.

Please open CD file 'ALPHA-2 (EMA PLUS) V1 & V2.' Note that the V2 calculation block is located starting at Col BM on the DATA sheet.

On the 'EMA 300T' sheet the violet EMA line follows the contours of the price trajectory quite closely but with greatly reduced 'jitter.' The 'smoothing' index obviously varies quite a lot. Note the classic valley at tick 2164 and the subsequent clean rise and reverse up-peak.

3. ALPHA-3 (THE LESHIK-CRALLE OSCILLATOR)

Motivation: Use an adaptive lookback on a tick series driven by recent %Range to provide the center for an envelope derived from the standard deviation with a lookback equal to the LMA standard deviation then calculate an oscillator to provide Buy and Sell trigger signals.

The oscillator uses a median or LMA-based lookback period whose range is automatically adaptive in the software between preset limits by calculating the %Range over a latest (50%) fraction of the median lookback. The default lookback is in the range between 300T and 600T for most stocks and needs to be set to the general activity of the stock.

We can also use the median instead of the mean as the median is more resilient to extreme values, so 'outliers' have therefore less influence making the algo more 'robust.'

The upper and lower bands are calculated (using 1.65 times the standard deviation) to provide a 90% likelihood for the envelope. Anything going up to or outside these limits is treated as a trading trigger. The steepness of the oscillator line is indicative to some extent of the quality of the signal.

The function does not appear to react symmetrically with most stocks as the down swing is usually more aggressive than the upswing but that is very much a feature of the individual stock characteristics.

Please go to your CD and open the file 'LESHIK-CRALLE OSCILLATOR.'

4. ALPHA-4 (HIGH FREQUENCY REAL-TIME MATRIX)

Basic Concept: This algo is currently still in development but we thought we would include it and give individual traders some basic ideas for how this could be implemented and a chance to do some development of their own.

The selected stocks must all be highly correlated (average 0.7 or more) on 200T % returns, say over two sessions.

The most basic grid uses four stocks (while we are hoping to use much larger grids these tend to require automation as there is too much information for a human trader to handle).

The 'curse of dimensionality' is operating here – so four stocks will need correlation pairs, AB, AC, AD, AE, BC, BD, BE, CD, CE, DE .

The divergence/convergence parameters are set using the %Range over the previous 300T as a baseline. An increase of 20% above the average baseline can be considered a trading signal.

Rising stocks are shorted, falling stocks are bought long when we are experiencing divergence. On the subsequent reversion the trades are closed out with a profit target of 35 basis points each.

5. ALPHA-5 (FIREDAWN)

Motivation: Build a surrogate for slope on the LMA to pick up long virtually continuous up trends.

This unusual algo is particular suitable to the energy sector where signals for it may be comparatively rare but when they do fire they tend to provide rather longer trades than we are normally looking for but with commensurably greater dollar returns.

The buy trigger is for the LMA (in an alternative version we use the EMA) to rise and exceed a return of 5 bp per 50T AND SMA > LMA for two consecutive 50T periods which gives us a TRUE condition for our Buy trade trigger.

The trade is held till the SMA crosses down through the LMA (SMA < LMA) constituting a Sell signal trigger.

Stocks reacting well to this algo are: OIH, RIG and other oil exploration and drilling-related stocks.

6. ALPHA-6 (GENERAL PAWN)

Basic Concept: Lead lag connection between stocks within sector and topological maps.

We have observed that some stocks are closely enough connected topologically that the movement of the bellwether stock (the 'General') is soon tracked by its neighbors (the 'Pawns').

The selection of the stocks may be helped by checking out the topology work of Professor Vandewalle. Also refer to www.impactopia.com.

Alternatively select a sector and use the largest market cap stock with a %Range greater than 0.0075 averaged over the last three sessions EOD as the General. Then select four more stocks from the sector with the largest %Range (also over the last three sessions EOD).

We set the bp trigger threshold for the General at 25 to 35 at 50T 'boxcar,' at which point, if TRUE the algorithm fires market orders for the 'pawns.' (As always, supported by the capital protection stops.) You do not have to go long the General unless you particularly want to. The concept is to use the lead/lag property.

There appears to be a 'time constant' analogy in the lag between the General and the Pawns. An oversimplified explanation would be that the Pawn lag is a reflection of the overall integrated market participants' reaction time needed to process the information and make the trade decision.

Using four Pawns it seems best to either put them all on trailing sell orders or to use a 30 bp profit stop.

7. THE LC ADAPTIVE CAPITAL PROTECTION STOP

The Concept: We let the very recent trade data provide the information for setting the LC Adaptive Capital Protection Stop.

In order to protect the capital invested in a trade (long or short) we must submit an opposite order to the market which is priced so that the current volatility regime for the traded stock does not 'whiplash' you out of the trade before it has a chance to develop or make the desired trajectory but at the same time protects you against a sharp contrary price movement.

The idea of 'stop loss' has always been set up as a basic ingredient to professional trading. In many cases individual traders do not put on protective stops as they are not sure how to set the value of the stop. The stop value often varies with the estimated holding period of the trade. Not using stops is, in our opinion, playing with fire and adding another layer of stress on the trader.

Setting plain stops as described in the literature by using a fixed percentage of the value of the trade does not work for the delicate high speed, high volatility trading that we use. The obvious reason is that the stops described in the literature are mostly for use with long holding periods whereas our protection stops are optimized for the shortest possible holding time which still provides us with our target return.

The LC Capital Protection Stop comes in two versions.

The mechanics of Version 1 are as follows:

1. Decide Lookback period based on current trade density per time. As we use ticks almost exclusively we will set the lookback as a tick value. Our usual default is 100 ticks back from the time stamp of our trade. This may need adjusting for the higher activity stocks.
2. Calculate the Standard Median Deviation (SMD) of the trade price over the lookback period. Here we are using the Median as it makes the standard deviation less sensitive to extreme values which would provide too much of a bias.
3. Subtract 2 SMD from the buy price (assuming we are using a long trade in this example).
4. Place a LIMIT SELL ORDER at the resulting value.

The mechanics of Version 2 are as follows:

1. Calculate the $Range on a lookback of 100T.
2. Subtract the $Range value from the buy price (assuming we are using a long trade in this example).
3. Place a LIMIT SELL ORDER at the resulting value.

All the parameters of the stop should be tested against the individual stock and current market conditions.

We use Version 2 more frequently than Version 1.

26
Parameters and How to Set Them

Having reached this far it is time to set up the trading parameters of the ALPHA-1 ALGO. (You may remember we defined a 'parameter' as a value in a formula in an algorithm which we set, or even which the algorithm sets for itself on occasion, clever thing that it is.)

We should stress at this point that all profitability in algorithmic trading in a very global way depends on the setting of the trading parameters.

We distinguish between two types of parameter:

1. Parameters which refer to the internal workings of the algo.
2. Parameters chosen as trigger points for Buy and Sell decisions.

We shall describe the setting of the Buy parameter for the ALPHA-1 ALGO. There are a number of ways of accomplishing this absolutely crucial step. We'll execute a short lookback (five sessions) parameter-setting method for this example.

We want to set a level for the trigger line so that its valleys reach or penetrate a straight horizontal line drawn over the five trading sessions a minimum of three times or more. If you find this cannot be accomplished retire the symbol to the 'holding corral' and choose another hopefully more cooperative stock.

Each one of these defined valley points is a trigger for a Buy signal. We head for the OMS and put on our trade – say 1000 shares of MSFT at Market. This will immediately fill and be displayed on your management system for you to monitor as it will display changes in price and the resultant open profit and loss on the trade.

As soon as your order fills put on the stop loss order from the stop loss trading template.

We now have 1000 of MSFT which we wish to sell at a profit.

Although this algo only has one buy setup we have several possibilities to close out the trade.

One way is to use a 'Dollar Stop' to trigger the Sell. (It should really be called the 'Basis Point Stop' but Dollar Stop is the commonly accepted term.) Monitor the basis point increase and make a judgment if you are satisfied with our standard target of say, 30 to 35 bps. If this has been hit or exceeded you cancel the stop loss and put on a Sell at Market order for the 1000 MSFT.

On certain stocks (and you need to experiment as this seems to be regime-sensitive and you need the 'now' state) the green trigger line can also provide very good Sell orders. This trigger fires from a peak, after the green line has traversed the red horizontal zero line. It works optimally only under conditions we have as yet not been able to define with adequate certainty, so there is a fair bit of empirical work that you may need to do to get really good results.

27
Technical Analysis (TA)

Technical Analysis (TA) can be defined as the discipline used to identify patterns in historical data with the conjecture that these will repeat, at least to some extent, in the future and afford us with trading opportunities.

Over the years many formulas have been developed by practitioners to quantify some of their ideas and provide more concrete bases for predicting price movement from historical data.

The art of TA has a very long history reaching back into the rice market in Japan from 1600 onwards. Kirkpatrick and Dahlquist give a very interesting historical background (at pp. 22–23, please see the Bibliography) of how a trader by the name of Kosaku Kato in Sakata City, a coastal city and center for rice distribution in Yamagata Prefecture, amassed a vast fortune trading by rules. This is the first recorded instance of a form of TA.

Though his methods were encapsulated in 'rules' known as the 'Sakata constitution' they analyzed today's prices to predict the next day's prices.

The USA start of organized trading is well documented – in the year 1792 Wall Street became recognized as the center for stock trading. A group of 24 merchants entered into an agreement called 'Buttonwood' as they used to meet under a Buttonwood tree. This created the first organized stock trading center. They named the organization the New York Stock and Exchange Board which was later renamed and is now known as the New York Stock Exchange.

In the USA Charles Dow is considered the 'father' of modern TA. He started writing occasional newsletters (which later became *The Wall Street Journal*) describing his theories and measures of the market in the late 1880s.

Here is a short review of a few of the basic methods with a view to providing a historical lead-up to algorithmic trading and possible inspiration for adaptation for design of trading algorithms, especially for tick-based, high resolution, high frequency systems.

Much of what you will read about TA may not be applicable to a straightforward 'proportional' adaptation to tick-based systems. TA is fundamentally based on daily interval 'homogenous' time series whereas our methodology works with 'inhomogenous' tick series (that means series where the time between data points varies – the series is thus asynchronous) and we use very short time periods, seconds, minutes, rarely an hour.

Some creative transformations may result in very usable strategies which would not have come to mind without having been 'triggered' by a TA paradigm.

To recap: The prime objective of TA is to predict where the price of an asset will move using historical data. It is now slowly gaining grudging academic acceptance and losing the 'chartist' stigma as the most modern computerized methods of analysis are now included under its banner. It has always found favor among practical traders and there is a very wide literature ranging from ultra-basic to extremely complex available to refer to. (Please see the Bibliography.)

An important point that cuts through much of the complexity: Trading strategies can be profitable only if the stock prices are either mean-reverting or the stock price is trending. This takes a bit of thinking about – you have to think about this in many resolutions for a start, then consider durations and frequencies, then imagine slopes and amplitudes. Its more than one latte's worth . . .

We will just skim over some TA formulas to give the algo designer some inspiration without going into the detail of how the particular TA is actually used by the practitioner.

Some TA can lean a bit heavily on statistics and this chapter assumes you have the Statistics Toolbox chapter down pat. However you will find that most of the material can be handled with plain vanilla arithmetic.

As we would like to use some of these properties in the design of high frequency ALPHA ALGOS we need an understanding of some of the basic TA now practised usually using a daily time resolution.

For those of you who want to delve deeper into this fascinating subject the Bibliography lists a number of books we have found especially useful. Please make sure to keep in mind that all the timescales are in days, whereas our timescales for most ALPHA ALGOS are in ticks (and thus minutes and seconds).

The adaptation process for us to use the current TA and adapt it for use with our algo designs is not trivial as we suspect that there are certain properties in the longer time frames which do not translate into ticks and seconds, just as certain human personality traits have different time constants. And it may be that some of these 'different' time constants are operational in some of the existing TA formulas.

CROSSING SIMPLE MOVING AVERAGES

Moving averages of all kinds are one of the main pillars of TA. The most basic and oldest being the interaction of short and long moving averages. When the short moving average crosses down through the long moving average (the so-called 'Black Cross') it's a signal to sell short or close out a long position. When the short moving average crosses up through the long moving average (the Golden 'Cross') it means go long.

Multiple moving averages may be used all the way up to the 'ribbon' method which uses ten progressively longer MAs. This takes a bit of getting used to in order not to be totally confused.

EXPONENTIAL MOVING AVERAGES

We have already previously described the EMA. Remember the earliest data values never completely disappear – it is simply that their influence on the current average fades the further back the earliest data is in the series.

BUY signals are triggered when a shorter moving average crosses up through the line of a longer one. It is advisable to use longish EMAs when trading a two-line crossover system.

We have had considerable success with single EMAs running on long tick look-backs, e.g. 200T. The valleys and the peaks on certain stocks were accurate to within 3 to 5T. Thus to trade these you Buy when the EMA hits a valley and then Sell when the EMA, having crossed the zero line, hits an up-peak. We have found the Sell side of this indicator rather unreliable and would advise using a $-Stop to close out the long trade.

By $-Stop we mean a predetermined profit, which, when it is reached activates a Sell at market order, for example 50 basis points (convert to dollars for the stock you are trading).

MOMENTUM and % MOMENTUM

Raw momentum is the difference of two prices over a defined interval.

Thus n-day momentum = price(now) – price(n) where (n) refers to the start of the period under consideration. For example n=3, means the closing price three days ago.

This is often also referred to as the Rate of Change (ROC) indicator. If the ROC is lower than the price today it may mean that the stock is running out of steam. It all needs interpretation and we prefer to divide the NOW price value by the start price thus turning the change into a percentage, making it much more useful.

RSI

J. Welles Wilder Jr created a number of TA systems which have lasted several decades. His seminal book *New Concepts in Technical Trading Systems* was published in 1978 by his own company. The book contains his original worksheets and a lot of his work was in the commodities sector. He was very aware of the concept of Momentum and its interpretation as acceleration and deceleration.

For example: 'The momentum factor (MF) is the difference between the closing price today and closing price two days ago. Go long on the close today when the momentum factor today is a higher number than the momentum factor for either of the previous two days.'

Wilder created the 'Relative Strength Index' (RSI) which has remained popular for several decades. This is the equation for calculating it:

$$\text{RSI} = 100 - (100/(1 + \text{RS}))$$

where RS = Avg 14 days closes up/Avg 14 days closes down.

The index itself will indicate tops and bottoms with values of 70 or higher (overbought) and 30 or lower (oversold), respectively. The index will top out or bottom out usually before the actual market tops out or bottoms out.

Wilder used a method for calculating moving averages which weights recent numbers more heavily – e.g. a 14-day Wilder moving average would be equal to the previous day's moving average times 13 plus the current closing price, all then divided by 14.

This method of calculating the moving average must be used with the RSI to avoid inconsistent results.

TRIX OSCILLATOR

This is a triple exponential average indicator for oversold and overbought markets oscillating about a center zero line. It may also be used for assessing momentum (positive values mean momentum is increasing). TRIX is calculated as the triple exponential moving average of the log of the price sequence. Long and short entries are signaled by crossing of the zero line.

PERCENTAGE PRICE OSCILLATOR (PPO)

A technical momentum indicator showing the relationship between two moving averages. To calculate the PPO, subtract the 26-day exponential moving average (EMA) from the nine-day EMA, and then divide this difference by the 26-day EMA. The end result is a percentage that tells the trader where the short-term average is relative to the longer-term average.

Calculated as:

$$\text{PPO} = 9\,\text{period EMA} - 26\,\text{period EMA}/26\,\text{period EMA}$$

THE STOCHASTIC

The word 'stochastic' actually means 'random' but the slightly misappropriated name for this indicator did not hamper its popularity. The story goes that the inventor gave it to a friend on the back of a newspaper and he had scribbled independently the word stochastic. His friend thought he meant it as the name. Anyway that's the story.

There are a number of interpretations of the formula. Here is ours:

$$\%K = 100 * (\text{today's close} - \text{12-day low})/(\text{12-day high low range})$$

The idea is to compare the Close with the range of a given lookback. The three default values are usually 26, 19 and 12 days, but a variety of others have been used.

Here's how to set it up in Excel for a 12-day period:

Col A, =MAX of previous 12 rows
Col B, =MIN of previous 12 rows
Col C, =MAX – MIN, this means Col A – Col B
Col F %K,raw =(today's close – MIN last 12)/(MAX – MIN)*100

Line one on the chart is called the %K with a default of 12.

Line two on the chart is a moving average of %K called %D with a default of five periods. The moving average may be simple or exponential and is usually displayed as a dotted line.

The Stochastic Oscillator always lies within the 0 to 100 range. 0 means the stock is trading at the lowest price of the 12-day period; 100 means it was trading at the peak of the prices set in the 12 days.

BOLLINGER BANDS

The standard Bollinger Band setup uses a 20-day EMA with two bands plotted above and below the moving average spaced normally 2 standard deviations apart. We have had success with slightly smaller separations of the bands, with a value of 1.65 Sigma which translates into 90% and gives us a few more trading opportunities than the 2 sigma value. The envelope varies with the volatility– when things are very hectic they widen, when quiet returns they narrow.

Buy when the upper Bollinger band is touched or penetrated. John Bollinger, the creator of this very popular study, notes that often there are sharp price changes when the bands narrow, signalling a reduction in volatility and possible reversal. Bollinger recommends 20 periods be used for the moving average. In live trading the trader often will make a value judgment looking at the rate of change and may trade before the top or bottom line are touched or penetrated.

WILLIAMS %R

This was developed by Larry Williams as a momentum indicator specially designed to flag oversold and overbought market conditions. Overbought means people are considering selling or about to sell. The opposite for oversold.

An unusual property of this indicator is that it seems to anticipate reversals of the stock price. It frequently will for a peak and it will turn down well before the stock price does the same.

It appears the property is based on something quite symmetric as it also turns up from a low before the price of the stock does the same.

The formula to calculate the oscillator Williams' %R

$$\%R = ((MAX_n - \text{Today's close})/(MAX_n - Min_n)) * 100$$

The oscillator swings from 0 the highest reading to –100 which is the lowest. Here 0 to –20 is considered overbought, and values from –80 to –100 considered oversold.

MOMENTUM (PLAIN VANILLA)

By tracking the amount of change in price over time we get some idea of the strength of the stock's trending behavior and to some extent can judge if this is holding, waning or in fact increasing.

$$MOM = \text{price now} - \text{price n periods ago}$$

This we often convert to percentages of basis points. Use a trend following oscillator and buy when it bottoms. Sell when it hits a peak and starts to turn down.

Often a strong trend will keep the indicator at the top for quite a long time so watch out not to miss selling when it starts to turn down.

THE ARMS INDEX OR TRIN

This venerable index was developed by Richard Arms in 1967 and has suffered a variety of names over the years. *Barron's* magazine first published it calling the Short Term Trading Index. It is often known as the TRIN (short for Trading Index).

We have included it here mainly because it has withstood the test of time and it is quite elegant in its calculation:

$$TRIN = (\text{advancing/declining stocks})/(\text{increasing/decreasing volume})$$

It shows if volume is flowing into advancing or declining stocks. A TRIN below 1 shows increased volume flowing into rising stocks, while a TRIN above 1 shows more volume going into declining stocks.

We have used it as a real-time market sentiment indicator while trading individual stocks.

The above short overview of TA studies and indicators only touches the surface of this discipline. There are over 2000 entries in the knowledge base of the Market Technicians Association.

We often study this discipline and try to adapt it to our high frequency trading requirements. We recommend that you do the same; browse the Bibliography, but most of all – experiment! We believe that only an empirical approach, well supported by the theory and the curiosity of an enthusiastic intellect, stands a chance in this hugely complex arena.

28 Heuristics, AI, Artificial Neural Networks and Other Avenues to be Explored

The word 'Heuristics' comes from the Greek 'heuriskein' meaning 'to discover.' It was introduced into England around 1800. The term then was used to refer to useful processes for dealing with problems that could not be handled by logic and probability. We can consider it as a 'rule of thumb' based on experience.

In 1905 the young Albert Einstein published his first paper in quantum physics: 'On a heuristic point of view concerning the generation and transformation of light.' In this Nobel Prize winning paper Einstein used the term to indicate that he considered the view he presented in the paper as incomplete but nevertheless useful.

A computer model of a heuristic needs to specify the steps of gathering information, the point at which information gathering is stopped, evaluation of alternatives takes place and a decision for a course of action is taken.

In our understanding of 'bounded rationality' the time and processing limitations of our brain or of any computing equipment, AI, or Artificial Neural Network are a finite constraint and the use of heuristics must be formally defined to be consistently useful.

In discussion we sometimes come across the idea of 'probabilistic' algos. We find that fundamentally confuses the issue – algorithms are by their very nature deterministic and from one set of inputs can produce only one set of outputs.

When we start dealing with probabilities in our view we enter the realm of 'Heuristics' which can be best defined as we mentioned above: 'rules of thumb' which usually give you the result you require, most of the time. But no guarantees.

The markets must be considered as complex, emerging, self-organizing adaptive systems and all algos face a moving and ever-changing target. Add to that the potential for regulatory changes and the picture is anything but simple to decipher.

It may seem a bit strange using a deterministic tool on such a moving target but on some reflection it actually makes sense; at least we know where the gun is pointing, precisely.

Perhaps we are being too pedantic as one could say that an algo which has a variety of branching tests is heading toward the less conceptually deterministic side. It's hard to really decide where probability starts taking a leading role, perhaps mediated by the number of outcomes one considers possible, or likely.

A new paradigm of 'structured heuristics' needs to be created with the goal of providing the 'requisite variety' to match the complexity of the markets with the adaptive flexibility to rapidly track the market evolution. Maybe 'Fuzzy ALPHA ALGOS' will come to our rescue.

This would mean dealing with the multidimensionality and the constant fluctuations in importance of various parameters in the understanding of the price trajectory of stocks. Unfortunately the famous 'curse of dimensionality' precludes a really comprehensive attack at the current level of development in the technology.

The current strategy seems to favor simplicity, to add as few dimensions as possible while still giving the required results. Occam's razor rules. This can be proved to have a greater chance of 'learning' from examples – we say it 'generalizes' to new data more effectively.

AI and Artificial Neural Networks (ANNs) when used for algo creation or to predict price movement in general have had a mixed history of success and failure.

Being computation intensive, with our requirement being real time, it was not feasible to implement sensibly until very recently.

We have spent about three years, starting in 2000, working in this area with little success, which we conveniently attributed to the inability of the hardware to keep up with the streaming data. The dataflow in 2010 is going to be multiples greater so we have to see if the computing hardware will win the race so we can use ANNs in 'almost' real time.

It is an area worth revisiting in the near future. It is tempting as here would be a possible implementation in line with the 'law of requisite variety.' Simply put – complex problems have complex answers. For those readers interested in adaptive systems and Artificial Neural Networks we would recommend a visit to www.peltarion.com.

29
How We Design a Trading Alpha Algo

Designing trading algos for the individual trader is our next target. The big battalions with their 600-man trading floors have their own techniques. We will describe our own methods.

Here is what Jane and I actually do when we start designing a new algorithm:

We always start with an assumption or conjecture about the price behavior of a target ticker symbol. This may appear, unbidden, over a cup of coffee or as a result of a vague specification requirement or from a 'wouldn't it be wonderful' bit of wishful thinking, or even be hidden in some ideas we have been kicking around for months but did not get around to looking at more seriously.

The inspiration may equally come while actually trading and looking at the tick charts developing, live.

Or we may start with a hoped for, or required, specification of what we would ideally like the algo to do. We kick the idea around and look at it from many perspectives, make guesses as to possible approaches, list 'brainstorm type' ideas.

We usually pick up a few charts of the target ticker symbol and also track its current live behavior. If some of our vague, half formed ideas look promising and it all starts to look interesting we then look at ten or more recent tick charts, with yesterday's tick chart on the top of the heap, and let the data whisper to us what secrets it holds.

We look at what moves with what, try to see patterns and similarities. We try to see if there is any structural similarity between the sessions.

We look over some of our metrics to see if anything jumps out at us as unusual, useful, surprising, not what we would have expected. We create time series from various metrics. Our favorites are the volatility, %Range and basis point returns areas.

At this point we often take a break – anything from a coffee to much more and let the whole thing simmer.

Sometimes a longer break overnight has produced good results. The mind has some sort of integration system which seems to work best at night when you are asleep.

Once we have pulled some ideas out of the ether and made a priority list we start defining things a bit more. We try to define clearly, in mathematical terms, what we are trying to achieve and how it relates to what we believe we see in the data. Careful here as to what we think we are seeing in the data – it may be only in our own mind as it is so programmed to make patterns of everything we see.

Again leaf through the tick charts. We are lucky in that Jane and I work together as one. We often finish each other's sentences ... Having quite different backgrounds we can also look at the problems from many different perspectives.

We now formalize a 'conjecture statement' on paper – an idea that can be clearly defined and write it out in pseudocode.

Pseudocode is our favorite method of defining the ideas and a basic tool in algo design. The first step in 'mathematizing' the conjectures and patterns.

It's a sort of natural language English which gives us freedom to express the logic without being constrained by a language syntax. The operators are similar to those you will find in any computer language. The flow of operations in any algo can be described by a limited number of constructs.

In our semi-structured pseudocode we also use the following computer language constructs:

SEQUENCE – one statement follows the previous statement.

So we are saying do these actions in sequence, one after another. This is indicated by writing the actions one line after another.

DO WHILE – keep doing this action as long as a certain condition is true. Also REPEAT UNTIL, this is the same but the test for the 'true' condition is at the end of the action.

IF... THEN gives us decision power – we make a branching choice between two courses of action. This can be extended to

IF ... THEN ... ELSE. If there is a choice to be made from more than three courses of action we use the CASE statement and make the selection from as many choices as needed.

FOR ... NEXT is our counter, if we want an action to be repeated a specified number of times.

We also use the Boolean operators AND, OR, NOT, XOR.

When we are happy that the script plays out what we intend, it is ready to be written up in function code (computer executable) and rough tested as an algorithm. We use, almost exclusively, Excel functions to interpret our pseudocode with rare excursions to Visual Basic or any of the more sophisticated languages such as JAVA. The preference for using the Excel native function language is based on the fact that we can get some fairly complex code to execute extremely quickly and efficiently as the Excel functions have been optimized in microcode and are thus very fast.

We also prefer a staged approach, especially when going into Excel function language – this is quite convenient and worth the effort. To make things as transparent as possible we split the initial algorithm functions statement over as many columns as required to make the flow totally clear and obvious. This catches errors in structure, logic and the actual language syntax very quickly.

Only after we are totally satisfied that we have instructed Excel to do exactly what we want it to do (this may mean running a fragment of the software to ensure it does what we think it should do) do we compress the function language to run in fewer columns. Often we leave this to a later stage.

Once the conjecture is 'mathematized' (converted into function language), we have an algo which a computer can execute and we are ready to have the first baby tests.

We then run the new algo on historical data and generate a set of metrics results for comparison with our set of 'desirables.' We may iterate a number of times playing with various parameters to see what effect these small changes have on the overall results.

We are getting ahead of ourselves. Let's jump back and briefly look at the list of things we would like our algo to achieve:

1. Provide Buy/Sell triggers for trade durations of no more than say 1000 seconds, with returns of not less than 40 basis points net per trade. (We will come to the prioritizing of trades using the Basis Points per Second method a bit later, as well as to capital preservation stops.)
2. We would like the algo parameterization to be stable over a reasonable time span or at least have it predictable from backtest over the last two to five trading sessions.
3. Over five sessions we look for an average of eight to ten trades per stock.
4. We aim for seven or eight wins against three or two stopped out trades.
5. Using real-time inputs to our Excel spreadsheet we want the algo to rapidly compute an output which tells us to put on a trade for the ticker symbol we are working on when the trigger conditions have been met. We want a 'trigger,' say, within 3 to 10 ticks of the ideal trigger point. We want the recalculation to not exceed 150 milliseconds.

Besides the requisite real-time feed with the ticker symbol, 1. date, 2. trade timestamp, 3. volume of shares traded and 4. trade price, let us look at what else we would like to have in the way of input materials to work with:

Up-to-date tick charts (five to ten sessions or better) created from the Excel templates are populated with symbol, date, time stamp, volume, trade price.

Excel templates for each ALPHA ALGO are configured but blank with no data.

Basic knowledge of the stock and its sector. (You will find most of what you need on two websites: www.marketwatch.com and finance.yahoo.com, but we will provide examples later which you can get started on.)

Barron's and the *Wall Street Journal* are useful to scan for relevant news and 'sentiment' (how the market is feeling – elated or depressed or 'normal').

Whatever books on technical analysis (TA) and statistics you may have read could be useful sources of inspiration. We have included a list of books we found useful in the 'Bibliography.'

Look through some TA books. Often a TA method which is used on longer timescales (like days) can be adapted to the tick paradigm or at least be a source of inspiration.

Look through recent news items on the financial websites. Become familiar with the stock. Peruse its history. Look at price movement as far back as you can, more for comfort and familiarity than immediate trading use.

Our conjecture is that each stock has a distinct 'personality' which is expressed by its trading characteristics. Get familiar with it. Get on speaking terms. Strike up a friendship.

The next thing to look at is what math and stats tools we need and are going to use.

We use a very limited range of tools, which is growing all the time to include creative combinations to exploit observable market stylistic traits such as 'pair trading' where we exploit the 'mean reversion' characteristics of highly correlated stocks. We will come back to that in a later chapter.

We have used a very small toolset to produce the algos which we describe here. Chapters 12 'Math Toolkit' and 13 'Statistics Toolbox' should provide you with all you need.

We have found that it is important that all participants in design and use of algos 'speak the same language' as the use of terms on Wall Street is often ambiguous and imprecise or both. In addition when we are innovating a new field with plenty of new and unfamiliar words there is an urgent need to define precisely new concepts and ideas for everybody to have the same understanding so we have included a chapter defining our 'Nomenclature' which is just a fancy way of saying 'this is what each word, symbol and acronym we use means in our little world.'

So let's dust off the keyboards and go through the design of an actual ALPHA TRADING algo together, step by step – live!

We have been thinking that markets trade 'sideways' for about 60% of the time (with the other 40% roughly split between up trends and down trends) and that during that time many stocks trade in a 'trading range.' This trading range has a 'ceiling,' which TA quants will call 'resistance,' and a 'floor,' which is called 'support.' Resistance because the price does not penetrate above this barrier and similarly with support, the price does not fall below it.

The idea is that we see the stock price gently bouncing up and down between these two barriers, at least for a while.

(When it eventually does penetrate it's called a 'breakout' or 'breakdown' but that is another story, another opportunity.)

So what would we have to do to take advantage of this 'trading range' pattern?

Jane asks how frequent the patterns are and do we know how much they vary from stock to stock and how long do they last and how far is it from the floor to the ceiling and how do we know when it will break?

Edward puzzles over this avalanche of questions for which the historical data hopefully will give us some answers. Now how do we have to set up the questions in a way that our computer will not choke on them?

Guess that we take the cop-out route and look at a few charts to get us going. Hey, there you go Edward – FCX (the copper people) on chart 33 spent half the afternoon in 'ping pong land' with the price trading between the resistance ceiling and the support floor.

We could say something like: look for alternate valleys and peaks separated by 300 basis points, or more, floor to ceiling, with the peaks not varying more than plus or minus 30 basis points between consecutive peak and similarly the consecutive valleys being within plus or minus 30 basis points of each other. The plus or minus values in bps are for illustration only and are not meant as a 'hard' value.

Now for the crunch instruction: if this pattern repeats (like you get three hits within say a period of 600 ticks), if the next swing hits the floor go long. (Buy at market.)

Wait for the stock to reverse, hit the ceiling, sell the long position and open a short position. When it retraces back to the floor, cover the short. And so on ...

OK, that sounds promising. Now to write the Excel function code for this little fella shouldn't take us long, looks simple enough (it took three days).

Besides the 'pure' strategies there is a plethora of 'second line' rules and parameters which can be overlaid on any of the algos.

The rules can range from terribly simple to extremely complex concatenated structures (sadly ignoring Friar Occam's sage advice). The more conditions the harder the generalization will be, but for short-term temporary measures we have found that this can work when used sparingly.

These, as always, are only limited by the designers' creativity and the algo design requirements. For example: 'Do not put on the trade if the S&P has fallen 3 points in the last 30 minutes.' Or: 'If the transaction frequency second derivative exceeds 'n' increase the volume of the trade by 50%.' Or: 'If stock A has been trading at twice the average volume and its price is falling 25 basis points per 600 seconds then do not put on trade for stock B.' And so on ad infinitum.

Placing of 'second line' rules can be permanent or just to cover a particular change in market conditions. Or even just a temporary band aid while problems are being resolved residing deeper in the algo's construction.

The construction of all algos can be further expanded and refined almost without limit by the use of the Boolean operators AND, OR, NOT, XOR and comparison operators EQUAL, GREATER THAN, SMALLER THAN as well as the conditional operators 'IF ... THEN,' 'IF ... THEN ... ELSE,' 'DO WHILE.' The palette is extensive and the possibilities are again only limited by the creative imagination of the designer.

Complex algo designs tend to have a shorter working life than simple designs.

Remember the cliché: LESS IS MORE.

30 From the Efficient Market Hypothesis to Prospect Theory

This chapter and the following two chapters (31 and 32) are included in order to give the new trader a mini review of the background thinking relating to the equity markets. They can safely be skipped by the reader until he or she is ready and eager to tackle some more complex concepts which will provide more background.

We believe that TA attempts to mathematize a reflection of universal human biases: fear, greed, uncertainty, expectation, impatience, envy. So we are going to also have a look at the work of Nobel Prize winner Professor Daniel Kahneman and his various colleagues working in the behavioral finance discipline. More about these follows after we have had a look at the Efficient Market Hypothesis.

The Efficient Market Hypothesis (EMH) assumes that investors and traders act rationally at all times and that information is equally and instantly distributed among them and is immediately reflected in the price of the stock.

This line of thought would make TA unusable and in the end would make trading itself impossible as all information would be priced into the market and there would be no incentive for traders to buy or sell as there could be no disagreement on price.

The 'strong' part of this theory has now been replaced by the 'semi-strong' and 'weak' forms which are still with us to some degree.

The research in the discipline of 'behavioral finance' has shown a large number of places where traders do not act rationally but follow a variety of biases inherent in the human makeup. These biases have been elucidated to a great extent by the seminal work of Professor Kahneman and his co-workers. (Please see the Bibliography.)

People tend to want to belong to groups which gives them a feeling of security and belonging and in general will conform to the group's opinions. Disagreeing risks rejection and possible expulsion from the group. So we have the kernel of 'crowd behavior' deeply ensconced in our basic makeup.

Not wishing to sell when all your friends and even CNBC news is of the opinion that you should sell puts you under stress.

We will dwell a bit further on the subject of 'biases' as it links people's emotions (rather than their so-called 'rational' behavior) with how they react to price movement.

We believe that the behavioral aspect of financial trading reaches into the deepest part of the human psyche.

Our mathematizing of constantly flowing mixtures of human emotions and desires, as it is developed into TA methods and formulas, is something we believe could be greatly improved. The problem with much of TA is that in our opinion it is much too rigid and 'coarse-grained' in its present implementation to capture the finer nuances of human behavior and thus works well only some, indeterminate, part of the time it is used, when 'all' things sort of match up.

We would need to encompass a much wider range of input dimensions – some of these would be very hard to mathematize in any parallel way with many of the other dimensions. Not an easy task.

Returning to the theoretical aspects of how TA is viewed by academia: The strict version of the EMH (Efficient Market Hypothesis) was developed by Eugene Fama in the early 1950s. It basically asserts that the market is driven by information and that all is reflected or 'discounted' in the share price.

By the 1990s the behavioral school had found quite a number of exceptions and anomalies in the theory and the originators of the EMH published an amendment: a weak form, a semi-strong form and a strong form of the hypothesis. Each of these has a different theoretical course in explaining the working of the market.

In all cases it makes the success of TA unlikely. Yet there are large numbers of practitioners paying no heed to the doom saying theory who have always used, and still use, TA in their trading.

Returning to the subject of 'biases' we must not forget that we ourselves are not immune and must be constantly aware of their presence and influence on our own behavior and take it into account.

We can easily 'see' relationships in data which are in fact false, or even random. Checking one's work has to be done with a good dose of self-discipline.

Trading in the 'Electronic Pit' is a prime example of decision making under uncertainty. You are facing an unknown number of market participants who are looking at exactly the same data that you are seeing on your screens. What are they going to do?

What will you do?

There is an extensive amount of research which indicates that people do not choose 'rationally,' in other words that their actions, when faced with a risky prospect, depart systematically from expected utility theory.

Research has shown that people show a significant propensity to selling winners too early and holding on to losers with the hope of the stock 'bouncing back,' until it is much too late to mitigate the loss incurred.

The claim that people do not behave according to utility theory is obviously troubling to economists whose basic assumptions are that people behave rationally at all times and that they choose exactly what they prefer in a transitive choice manner.

However it now appears that preferences in decision making are particularly labile and are influenced by how the decision choice is presented and numerous other factors.

Traders can be swayed by incoming news, discussion with colleagues and the movement of the market itself. The personal decision process is fragile. Overreaction can be triggered by emotions. Positive feedback of sufficient magnitude, persistence and speed can create the 'bubble to crash' syndrome.

We know that supply and demand establish the price of a stock. When price does not change you have to look at the volume being transacted to see what is really happening. If the volume is relatively high you could interpret that as a strong 'difference of opinion' in the market participant factions. The number of buyers and sellers is always equal. (Every trade must have a counterparty!) One trading cohort feels that $45 is a cheap price to buy WIDG and the other trading cohort is quite certain that $45 for WIDG is a great price to get, so sell! And so the difference of opinion creates the trade, one side buys and the counterparty sells.

A stronger 'difference of opinion' is seen when the price starts moving in a trend. This tells us there is an imbalance in the supply and demand for the stock among the participants – this 'difference of opinion' may be driven by rational information and its interpretation by some of the participants, or it may be driven by fear of being left out of the party (assuming this is an up trend), greed as they see profits, and so on.

The steeper the price change slope of the trend the greater is the 'difference of opinion.' The amount of shares being transacted is another signal which needs careful watching as it is the second metric with which we can gauge the strength and 'health' of a trend – a marked decrease in volume may signal that the trend has just about run its course. One also needs to keep an eye out for transaction frequency, the number of transactions per second, or minute.

Motivation for understanding the various dimensions of TA coupled with some further understanding of behavioral theories is the hope that these will provide us with some clues of how to design our ALPHA ALGOS.

Let us start by reviewing some of the work of Kahneman, Tversky and others, much of which was subsumed under the title 'Prospect Theory.' (Please see *Prospect Theory: An Analysis of Decision Under Risk* (1979) in the Bibliography.)

Don't let the date put you off – it just shows that this discipline has been going for some time. Besides, Kahneman only got his Nobel Prize in 2002, rather late in the day.

The core of the theory states that when faced by choices among risky and uncertain outcomes prospects show effects that are totally inconsistent with basic utility theory.

In our language it translates as: 'We do not act rationally under conditions of risk and uncertainty.'

This causes a major problem for the economists whose main precept is that we all act 'rationally' all the time. This kind of 'simplifying assumption' is particularly invidious as, if it is not 'flagged,' it passes as the normal course, which is flagrantly inaccurate.

Even if we were all rational all the time the diffusion of the information on which we took our decisions cannot be instant, nor equitemporally distributed nor equally assimilated. In addition each market participant will have a different interpretation as well as different stressors to mould his trading decisions and their timing.

In prospect theory, loss aversion refers to people's tendency to strongly prefer avoiding losses to acquiring gains. Some studies suggest that losses are twice as powerful, psychologically, as gains. Loss aversion asymmetry was first convincingly demonstrated by Amos Tversky and Daniel Kahneman.

Now for more of the actual biases the research has uncovered. We prefer to call them heuristics as they are something more akin to a rule of thumb.

One of the very first of these has the tongue twisting name: 'representativeness.' (You get used to it after the first 10 tries . . .)

In order to survive humans have developed by evolution a quick way of classifying objects, people, events and even their own thought processes. This works OK most of the time except when a new class of these items comes on the scene and does not quite fit any that are currently available.

We then tend to force fit it to what we think is the closest fit. That is fault number one – possibly it is not even in our repertoire at all and we will misclassify it with all the errors that will then follow.

As we are busily working what is most similar (representativeness) we tend to neglect a couple of vital concepts: 1. We become insensitive to the base rate frequency of prior outcomes, and 2. We neglect to consider the sample size. People misjudge the likelihood of a short sequence of events providing the same outcomes as a much longer one. This is called misjudgment by the 'law of small numbers.'

More representativeness (you can say it now, right?) can be observed when you get a number of similar and redundant and highly correlated variables. Confidence in the choice of action zooms! Wow, it all resembles what I already know! Alas not for good reason as redundancy among inputs will severely decrease the accuracy of the decision even as the subject's confidence in being 'right' is increased. (This may be helpful in deciding that the shadow under the palm tree is in fact a tiger and that we should immediately run like the blazes, (which we do) while our data-prone proof-seeking colleague goes to investigate the shadow which indeed turns out to be a tiger and possibly gets eaten, unless, hopefully, he is an Olympic 100-meter sprinter . . .)

Another nuance is that people tend to perceive things in the light of their personal experience (which may or may not be relevant) while the statistics point in another direction (which they will duly tend to ignore).

The next heuristic we must look at is 'availability' (this is a bit easier on the tongue). Here people assess the frequency of occurrence by the ease with which occurrences of this class can be rapidly brought to mind.

You see a particularly gory pileup on the turnpike and you drive more carefully for the next few days, till the memory wears off.

'Anchoring and Adjustment' are another heuristic bias we are all prone to. This happens only when you have, or are given, a starting point. Like yesterday's close of Intel was ... One tends to weight this too heavily and 'adjustment' away from the anchor is not easily accomplished. We all have an innate bias towards not straying far from our initial estimate even when confronted with very convincing but conflicting evidence.

'Belief in the law of small numbers' sounds so tempting and is so very fallacious. Most people feel that random sampling will satisfy the law of small numbers just as it also satisfies the law of large numbers. It doesn't! The law of large numbers practically guarantees that very large samples are highly representative of the population from which they have been randomly selected. Small numbers do not yield that property.

Richard Thaler has added a few more concepts: We will look at The 'Endowment Effect,' and the 'status quo bias.'

You can see 'endowment bias' in operation when people do not sell stocks which they have held for a while. Actually just the fact of ownership of a stock immediately 'endows' it with some added psychological value.

The 'status quo bias' is a cognitive bias for the status quo; in other words, people tend not to change an established behavior unless the incentive to change is most attractive. There appears a certain amount of inertia in the requirement to change direction.

Our interest in this discipline of behavioral biases obviously stems from a desire to help us understand our analyses of historical tick data and also to elucidate some background for the construction of new ALPHA ALGOS.

We perceive a rather fundamental problem in that the data may need temporal stratification as the long runs must reflect changes in the ecology of the market, the regulatory framework, the long-range business cycles and possibly even the ultra-long range but not completely researched cycles of Kondratieff and other long-range cycles. Much of the work done by the Tier 1 companies often uses extremely long lookbacks.

An algorithm that would work 'from the beginning of time' is obviously a Holy Grail as it would be enormously robust and produce profits on a nice predictable and regular basis. We have not found any algos that fit the description but we have not stopped looking yet. Most of our work is using comparatively short lookbacks, from five to 60 trading sessions.

To give you an idea of the magnitude of the task: In order to analyze 100 stocks by looking at tick data for 60 sessions requires a total of 6000 Excel files. With each file analyzing a variety of metrics, some converted into series, it is not a trivial exercise.

This rather dispassionate academic approach is all overlaid on the basic (or shall we say base) emotions humans are heir to: greed, fear, regret, envy and more which influence trading decisions.

The entire basis of the Leshik-Cralle methodology is 'local' and short lookback. Volatility regime shifts in this context become just large perturbations among the general turbulence. Our high resolution of tick data should provide us with as clear a picture of the price movement as is feasible using the data inputs with which we are working.

Therefore, in our day-to-day ALPHA ALGO development, we use a very local approach as we believe that the area of influence we are looking at is comparatively small most of the time.

The confrontation is between a deterministic approach and a probabilistic environment. The market dislikes ambiguity and uncertainty of outcomes – it becomes nervous and skittish, and in the final analysis totally unpredictable as emotional feedback swamps the rational controls.

Our attempt at visualizing the market system could take many pages. Here we shall try in a paragraph:

We see a system having a myriad driving forces: some very stable, some quite constant, and others which vie for precedence and exchange leadership roles in a complex interrelated set of rules which we are not privy to. New forces come into play, some stay and get stronger and last a long time while some quickly wither away; others have a crescendo driven by some unknown force, then suddenly disappear. Some only make a rare appearance. Some stalwarts are there all the time, sometimes leading sometimes following, sometimes stronger and sometimes weaker, but always there. Some feed on others and get stronger, others fade into nothingness.

It is quite a task to visualize.

31
The Road to Chaos (or Nonlinear Science)

This chapter, just like the previous chapter (30) and the following chapter (32) are included in order to give the new trader a mini review of the background thinking relating to the equity markets. They can safely be skipped by the reader until he or she is ready and eager to tackle some more complex concepts which will provide more background.

As our introduction to the complexities of the markets was initiated by the work of Professors Doyne Farmer and Norman Packard and kept humming by the incredible corpus of work of Professor Benoit Mandelbrot and the 'first hand' storytelling of Professor David Ruelle and the Santa Fe Institute, we feel we must touch upon some of their work more than just by a mention in the Bibliography.

The mathematical term 'chaos' was coined in a paper by James A. Yorke, Professor of Mathematics and Physics and chair of the Mathematics Department at the University of Maryland at College Park, and T.Y. Li in 1975.

We are delving into this rather difficult area as we are thoroughly convinced that in order to compete in the design space for strategic trading algorithms there will be a number of (as yet unknown) paradigm shifts which the designer will have to absorb and master. Being prepared for these shifts is half the battle.

Let us start with some terminology definitions so that we can talk about them.

'Phase space' – this is a purely mathematical construct where the coordinates represent the variables that specify the 'phase' or 'state' of a dynamical system at any point in time.

To the word 'dynamical' we will assign the meaning 'changing with time.' We can thus say:

'Dynamical systems' are anything that moves in time, or varies with time, anything that evolves in time. That is, any system where each successive state is a function of the previous state.

Most of chaos theory deals with phase space, and the machinations one can play within it.

'Chaotic' systems have 'attractors' in phase space that over time (and perhaps many, perhaps myriads of iterations) will tend to attract all the motion of the orbits coming from the same system's starting conditions (the 'basin of attraction'). No matter how turbulent the system it will eventually settle down to the attractor which has its long-run trajectory in its evolutionary path.

Edward Lorenz first plotted his three differential equations for weather forecasting and got the now famous 'strange' looking owl shape on his chart – this is where the 'strange attractor' got its name.

Much of the research work has been in 'lagging' the phase space which is a common tool in time series analysis. This leads us to the 'embedding dimension' and reconstruction of the trajectory of the system under study. When used for prediction we would use a succession of lagged values to predict the future values in the lagged series. (See Floris Takens in the Bibliography.)

'Chaos' in the mathematical use of the word is not usually random but is deterministic.

Going back as far as Henri Poincaré (who unfortunately did not have the technological means to demonstrate his ideas) the concept that even the simplest equations and ideas can lead to vast complexity was a concept with which he was very familiar.

The concept of turbulence (David Ruelle and Floris Takens, On the nature of turbulence, *Communications of Mathematical Physics* 20 (1971)) had a rough time getting published in a refereed journal and Ruelle had to cut some corners to get it seen (he arranged for it to be published in a journal of which he was the editor . . .). The scientific establishment took a while: 'first a few physicists thought these were controversial ideas, progressively they became interesting ideas, then well-known ideas.' (Guess the ultimate insult would finally be 'Oh, we knew it all the time . . .')

Turbulence has many analogies in financial series and Floris Takens was probably the first to suggest 'embedding' to reconstruct the trajectories of dynamical systems.

His work was cited by Farmer and Packard in their original Chaos paper.

Mitchell Feigenbaum discovered that when one changes the forces acting on a physical dynamical system we often see 'period doubling.' The time it takes to make a complete circuit (which is called the 'period') doubles.

The story goes that he (in the mid-1970s while at the Los Alamos National Laboratory in New Mexico) was playing one night with his computer and noticed the frequency of a dripping tap in his kitchen, and that by careful tap adjustment he could get it to double the frequency.

Looking at the logistic equation (it is shown below) behavior for $3<k<3.57$, the period doubling range, the transition in system behavior is induced by the tiniest change in the value of 'k.' The strange 'universality' thing is that all systems which undergo period doubling share this behavior both qualitatively and quantitatively.

This discovery of universality in the raw was a great advance in understanding the onset of chaos and nonlinear science in general. In real numbers Feigenbaum found that k values for each new periodicity came about at an approximately constant rate, which at the limit leads to his constant below.

Feigenbaum found this to be a universal property of a whole class of functions and discovered the constant which has the same transcendental properties as 'pi' and 'e.' (It still has to be proved to be truly transcendental.) The Feigenbaum constant = 4.6692 . . .

Going onwards, the very innocuous looking and very celebrated 'logistic equation' (it's also, more correctly called a 'recursive map') is:

$$X_{n+1} = \alpha X_n(1 - X_n)$$

for the interval $0 \leq X \leq 1$

with the control parameter $0 \leq \alpha \leq 4$ ($\alpha = k$)

Beyond a certain value (as we mentioned above) the sequence becomes more irregular and finally chaotic. (There is a good animation on Wikipedia under 'Logistic map.')

This discovery led to a whole area of investigations and as Professor Farmer said (in Fisher 1985) 'It's the universality that's exciting. It says that all nonlinear equations that do this particular thing do it in the same way.' And this could be in electrical circuits, optical systems, solid state devices, business cycles, populations and learning (Briggs & Peat, 1989: p. 64) and in our view also relates conceptually to the development of price discovery in the markets.

(Just by the way, scientific journals rejected Feigenbaum's papers for over two years before finally accepting some . . . this is not unusual as the 'establishment' referees are notoriously known to err on the side of caution. Remember the classic story of the Ruelle/Takens paper?)

With this little experiment under our belt just imagine what the electronic pit is up to, how many traders are looking at the same screen, how many are in desperate need of a cup of coffee and still feel queasy from yesterday's binge, how many had a row with their spouse, how many have to get home early for Aunt Julia's birthday party, how many have the mortgage payment looming on their mind, how many did not get a decent night's sleep, how many are bored stiff, how many like the girl next door, how many have a sick pet kitten at home called Simon (Simon is a girl).

Chaotic systems are very sensitive to 'initial conditions,' their state at the start of play.

The slightest difference at the start magnifies out to the system taking a completely different trajectory. This trajectory is not predictable exactly but in chaotic systems tends to revisit a 'basin of attraction,' never actually repeating the same path. When we plot some of these systems they often trace out areas which are called 'strange attractors' to which they are wedded. (Please see our Bibliography.) Why 'strange'? Well, they look strange don't they? The very first one ever looks like an owl.

The markets are chaotic and complex so we must accept the challenge. The markets have been going only a twinkling in cosmic time and the overall systems are dependent on time evolution with sensitive dependence on initial conditions, on top of which the whole system is driven by 'intelligent' market participants. If we wish to draw parallels with physical systems it is important not to forget compared to our trading universe that these physical systems are comprised of a many orders greater number of 'participants' (although perhaps not as 'intelligent') and super precise initial conditions which in any case we have no way of even thinking of ever retrieving. Overlay on all this massive regime, regulatory and structural shifts and things get pretty tough. Then add the behavioral concepts.

It was only in 1975 that Professor Mandelbrot coined the word 'fractal,' creating a new geometry of nature which made it possible to express such concepts as 'roughness' quantitatively. Having studied the early cotton price records he demonstrated that these were not truly Gaussian and created a backlash resonating even today. We shall take up the response challenge in a future volume.

Professor Mandelbrot has pointed to the fact that price variation is not the same for all markets and that a single statistical model might not describe all markets. Jack Schwager, as I recall, quite a long time ago was of a slightly different opinion – that all markets share the same characteristics.

Though one would have to agree with Schwager that there are quite a lot of shared characteristics, we think there are also quite a number of individual characteristics which are specific only to particular markets and, on a finer scale, even specific stocks.

It may be that there are many market properties which have similar characteristics but equally we have found quite a collection of variables which appear to be stock-specific and are potentially powerful generators of algorithmic strategies. We take the view that even individual stocks are a law unto themselves.

That the markets are man's most complex achievement/challenge is a pretty much undisputed fact. The fact that their dynamics are chaotic or verge on chaos is also unarguable

We have dwelt on chaos not so much as a *potential* source for algorithmic strategies (though there are some points which may be directly useful in the future) but rather as a conceptual tool to weave the thoughts of great complexity and delicate and unpredictable outcomes from what look like simple constructs into our thinking and creative processes.

Chaos and complexity are here to stay. (Actually they have always been here . . .)

As Klaus Mainzer so beautifully quotes Immanuel Kant at the end of his book, 'Thinking in Complexity': What can I know? What must I do? What may I hope?

32
Complexity Economics

This chapter and the previous two chapters (31 and 32) are included in order to give the new trader a mini review of the background thinking relating to the equity markets. They can safely be skipped by the reader until he or she is ready and eager to tackle some more complex concepts which will provide more background.

Currently there is no widely accepted definition of 'complexity' as it relates to the economy. This entire field still needs the 'under construction' sign.

With thanks to the brilliant minds at the Santa Fe Institute in New Mexico, Brian Arthur, Kenneth Arrow, Steven N. Durlauf, David Lane, Stu Kauffman, Doyne Farmer, Norman Packard and many others, we will paraphrase some of the thinking regarding the behavior of complex systems relating to economics.

Complexity economics rejects many features of traditional economic theory. The models used by traditional economics were all formulated as analogies to early models of thermodynamics, mainly based on the First Law of Thermodynamics – equilibrium. No free lunch, matter cannot be created nor destroyed, only transformed (with a small apology to Albert Einstein – the total stays constant even with $e = mc^2$).

Sadi Carnot established the principle that eventually became the Second Law around 1824. Rudolf Clausius formalized it as the Second Law of Thermodynamics in 1860 – 'In any spontaneous physical process entropy increases,' or 'The entropy of an isolated system not in equilibrium will tend to increase over time, approaching a maximum value at equilibrium.' Or 'Entropy grows in any spontaneous physical process.' Or, most popular, 'Heat flows from the warm body to the colder body' – there are a multiplicity of definitions.

Complexity economics claim that traditional economics has not adapted to the Second Law and thus remains an incomplete model of reality at the very roots. In addition information entropy has not been properly introduced into mainstream traditional economic theory. Information entropy was developed in 1949 by C. Shannon and W. Weaver, based on Boltzmann's statistical thermodynamics, as 'information uncertainty,' associated with any probability distribution.

Entropy has been used at least since 1988 to formulate the important concepts of organization and disorder, viewed as basic state parameters, in describing/simulating the evolution of complex systems (including economic systems).

In the light of the new concepts introduced, economic systems shall no more be considered as 'naturally' inclined to achieve equilibrium states. On the contrary, economic systems – like most complex and self-organized systems – are intrinsically evolutionary systems, which tend to develop, prevailingly toward levels of higher internal organization; though the possibility of involution processes – or even of catastrophic events – has to be taken into account.

Traditional economic models have been constructed by allowing only a very small amount of degrees of freedom, in order to simplify models. The ploy of 'simplifying assumptions' has served science well but it has also created a huge pitfall for the unwary and the uninformed.

To illustrate, the relation of unemployment and inflation is traditionally considered to be a simple function with one degree of freedom, allowing for very little entropy. So much for simplicity . . .

As to the practicability of theoretical instruments, there is also a crucial difference to allow for: traditional economics was conceived before computers had been invented . . .

From the Santa Fe Cabal (with thanks and admiration) we have gleaned a few more ideas which we have paraphrased below:

Continual Adaptation – Behaviors, actions, strategies and products are revised continually as the individual agents accumulate experience – the system constantly adapts. The system can be defined as all the human agents involved in the trading process, rules and regulations by which the trading process is governed and the physical layer of connectivity and computer processing systems.

Perpetual Novelty Niches – These are continually created by new markets, new technologies, new behaviors, new institutions. The very act of filling a niche may provide new niches. The result is ongoing, perpetual novelty. The participating agents constantly seek a trading advantage. New participating agents enter the fray, others retreat to safer waters.

Out-of-Equilibrium Dynamics – Because new niches, new potentials, new possibilities, are continually created, the economy as a whole operates far from any optimum or global equilibrium. Improvements are always possible and indeed occur regularly, possibly continuously. Improvements are the natural state and should be encouraged.

Information entropy was developed in 1949 by C. Shannon and W. Weaver, based on Boltzmann's statistical thermodynamics, as 'information uncertainty,' associated with any probability distribution. Yet most mainstream economists are yet to introduce information entropy to their models.

Entropy has been used at least since 1988 to formulate the important concepts of organization and disorder, viewed as basic state parameters, in describing/simulating the evolution of complex systems (including economic systems).

As previously mentioned, in the light of the new concepts introduced, economic systems shall no more be considered as 'naturally' inclined to achieve equilibrium states.

Economic systems – like most complex and self-organized systems – are intrinsically evolutionary systems, which tend to develop, *prevailingly toward levels of higher internal organization*; though the possibility of involution processes – or even of catastrophic events – remains immanent.

Traditional economic models have been constructed by allowing only a very small amount of degrees of freedom, in order to simplify models. For example, the relation of unemployment and inflation is traditionally considered to be a simple function with one degree of freedom, allowing for very little entropy.

As to the practicability of theoretical instruments, there is also a crucial difference to allow for: as previously mentioned traditional economics was conceived before computers had been invented. Computational simulations have made it possible to demonstrate macro-level rules using only micro-level behaviors without assuming idealized market participants.

Complexity economics is built on foundations which draw inspiration from areas such as behavioral economics, institutional economics and evolutionary economics and in some cases statistical mechanics. It is often said that Wall Street suffers from a bad case of 'physics envy.' Not enough really as one should add a proper dose of 'biology' envy to the mix. Then we should stir in the multidimensionality of the mathematics (still well out of reach even for the most powerful machines) and finally add a sprinkling of the frailties we humans are heir to – greed, envy, fear for a start. Stir well.

Complexity incorporates components from each of these areas of economic thought, though the true complex perspective must include many more characteristics to describe a dynamic system, such as emergence, and sensitive dependence on initial conditions. Complex systems usually incorporate a selection mechanism as described by most general evolutionary models. There is no widely accepted specific definition for complexity as it pertains to economic systems.

This is largely because the field as a whole is still 'under construction.'

33 Brokerages

In order to 'pattern trade' (this is the SEC's name for what we would call a frequent trader or day trader) you must open a brokerage account with a minimum of $25 000.

In practical terms this really means $50 000 or so to give you some slack. As is always advised by everybody, it is best not to use money that would change your lifestyle if it went south.

The various brokerages provide very different trading platforms with quite a variety of functionality so you need to look at a few, and, we would advise, perhaps try their demo packages before deciding which one you want to start with. Take your time as this is one of the more important decisions you have to make.

Many brokerages provide extensive toolkits and add-ons, e.g. TDAmeritrade provides a wide variety of useful tools.

Platforms such 'Lightspeed,' 'Omnipro' and 'Rediplus' are all available from the TradePro brokerage at www.thetradepros.com and are worth investigating and having a serious look at.

We find Townsend Analytics' 'Realtick' Order Management and Execution platform very easy to use but obviously it is largely a matter of personal preferences. We have also found their support staff always very helpful.

Stuart Townsend and his wife Margwen were the pioneers of direct access trading in the early 1990s and have kept their products as industry standards ever since, in the face of enormous competition from brokerages, banks and even Exchanges.

Their contact number for professional services and RealTick support or any enquiries regarding their products and services is:

Monday through Friday
7:00 am to 5:00 pm CT
(800) 997.9630
(312) 442.8970
support@realtick.com

The list of direct access brokerages which follows all provide a version of the RealTick platform:

www.eoption.com
www.Investscape.net
www.hamptonsecurities.com
www.lightspeed.com
www.mastertrader.com
www.nwtfinancialgroup.com
www.tnfg.com Terra Nova Financial LLC
www.Tradewithvision.com.

Other major brokerages with global reach (not a comprehensive list as it includes only those companies which in our personal opinion are of interest):

www. Interactivebrokers.com
www.tradestation.com
http://gset.gs.com/rediplus Goldman Sachs ElectronicTrading
www.tdameritrade.com.

The commission costs per trade need to be looked at. We find the 'offers' sometimes a bit confusing. So do hunt around for the best deal as long as the functionality you require is provided by the brokerage.

The only good way we know to see if a particular trading platform suits you is to try it out. This may be tedious and time consuming but we have not found a better way to accomplish this selection task. Most brokerages will be only too happy to provide you with a test account.

34
Order Management Platforms and Order Execution Systems

These two terms have been the subject of much discussion in the financial press. As many of them overlap with each other to some extent it needs a careful look at what a particular platform or system offers.

In the real world a trader could not care less what name we give it as long as it does what he needs so we can consider the OMP, OES and other versions of the acronyms as functionally much the same.

In addition it appears that the trend is for the two to merge into one by the simple process of evolutionary survival of the fittest.

In theory the Order Management Platform gives the trader information as to ticker symbols selected, real-time Bid and Ask, Time and Sales (streaming transactions as they are reported by the Exchanges) and the ability to select a wide variety of order types. We only use Market and Limit orders.

Some platforms will also show orders waiting execution, their part by part fills with the respective prices and an average price for the trade. Usually you get a list of live trades with the open P&L as well as closed out trades with their P&L. Much of this information is usually written to a log which you can access and download for archive reference.

The bare bones execution system is supposed to find the best routing for your order. We found little to recommend using a special routing system as our trade size does not warrant it. We often specify the routing together with placing the order on the OMS and very often will go to an ECN (ECN stands for Electronic Communications Network). When directing your trade to execute on an ECN you are going to have Buy and Sell orders matched automatically at your limit price. If you mark the order ECN only then you may be restricting your liquidity but this is not an issue when we are trading 1000 to 2500-share orders.

35
Data Feed Vendors, Real-Time, Historical

Practically all brokerage trading platforms include a data feed which you can access and feed into Excel spreadsheets, usually with relative ease. Query the technical support team on feed handlers to get the real-time data into your Excel sheets. Do the same if you are investigating buying the tick data from an independent data vendor.

The main problem is that most brokerages usually give only limited historical access, say five sessions' worth. This is not enough for our purposes of stock personality definition and clustering.

Tick historical data is available for NASDAQ and NYSE from a large variety of vendors. There is a list on nasdaqtrader.com.

If you decide to purchase historic tick data we suggest that a lookback of 60 trading sessions (three months) is more than adequate. We did not find longer lookbacks particularly useful at this time when dealing with fast tick-oriented trades.

Longer lookbacks could be of interest when we attempt to uncover very long-term patterns, deeply buried in the human psyche and its interactions with the markets, which may perhaps return on some sort of longish cycle.

We do not have any doubts that long-term patterns could be profitably researched but we think this will require a totally different approach and probably orders of magnitude more data than we are using at the moment. It is likely that many more variables might have to be considered but that leads us into completely different research areas requiring considerable fire-power from both human and machine.

Once you have purchased the data, as close to up-to-date as you can, it will probably be provided on an external hard disk. (DO NOT agree to a block download over the Internet as the quantity of data is much too large and the likelihood of errors too severe.) It is best to set up a file for each stock symbol and copy the data to it from the media the vendor sends you. We suggest using an external 750GB hard disk to archive all data.

From then on it's down to you to copy the day's data into your database and keep it current. This is no mean feat of discipline to do by hand. (Automating this sounds like a good idea.)

As mentioned above, it is vital when you are selecting your data vendor to make sure that they will support linking the real-time feed to Excel spreadsheets and will help with the 'feed handler' software. Allow yourself plenty of time to get everything working properly.

36
Connectivity

The issue of connectivity is often neglected by individual traders and yet it is a key to successful trading.

Much depends on where you are located and you will have to do local research as to what is available and represents value for money. Tell the vendors what you need the line for and make sure that you can get a minimum of 18 Megabits per second bandwidth, uncontended, if possible.

Look out for the 'contention trap' where you effectively share the connectivity with others. The more people log on the less bandwidth you are left with. Usually not a problem but we advise you to check with the provider to make sure.

There are a number of speed check programs which you can download from the net to see if you are really getting what you are paying for.

37 Hardware Specification Examples

The issue of what hardware to use is fraught with trade-offs within trade-offs. Good examples of 'recursion.'

If the budget can stand it we would recommend getting the most powerful box, with as much RAM as it will take. Say two quad processor boards, minimum 16 gigabyte RAM.

Best current NVIDIA card that will drive up to eight monitors.

Investigate solid state hard disks, they are new and expensive but if they live up to advance billing they are certainly worth it. (Our next rig is going to be fitted with two solid state hard drives.)

As far as monitors are concerned, it's all a matter of taste. We prefer an eight-way array (2 × 4) of 21″ with 1600 res min. (Samsung or LG). Larger monitors are preferred by many traders.

Operating system Windows XP Pro with the latest updates.

Suggest both standard keyboard and a glidepad ('Glidepoint Touchpad Mouse,' SmartCat or SmartcatPros) from www.cirque.com in Utah. Edward finds the glidepad (plugs into USB port) close to indispensable, says he could not work without them, and wonders why other people just don't get it. Jane says 'No comment.'

The hardware schedule which we use in our next rig is sketched out below as an example with approximate price indications on major items.

WE-5502-XG8 Stratosphere Elite 5502 XG4GB DDR3 1333 Unbuffered Memory $5800

1 CPU-Xeon X5560 Intel Xeon X5560 Nehalem 2.8GHz Quad-Core Processor $2600

1 UPG-HDD-SSD-64G Intel 64GB Solid State Drive, X25-E Extreme $700

1 UPG-HDD-640-BB 640GB WDC bootable backup drive + 640GB secondary drive $125.00

*Secondary drive set as a bootable mirror, zero down time
Quiet 22X dual-layer DVD±RW burners
Microsoft Windows XP Professional w/Service Pack 3
Total digital DVI monitors supported: 8
Beyond Rocket Science™ Service Commitment
1 SW-Office 2007 Pro Microsoft Office Professional 2007 $400.
ZQ-HB24HS Zenview Quad Elite
24″ super-premium S-IPS LCD panels
3840 by 2400 total screen resolution
1920-by-1200 WUXGA screen resolution per panel
1000:1 contrast ratio; 400 cd/m^2 brightness
178° (H)/178° (V) viewing angle
Ultra-fast 6ms response time
3-year on-site service/advance replacement warranty
$7200
SW-Zenview Manager
Helps you unleash the full power and productivity of multiple monitors.
Smart Taskbar, adds taskbars on all monitors
Move Window, two click monitor to monitor window movement
Application Positioning, opening windows at specific locations
Multi-monitor wallpaper, control the desktop of each monitor
... and more
CBD Combo Discount (5%) $816

The total, including various sundry items, comes to nearly $17 000 which sounds steep but one only gets what one pays for, and this is a top of the line professional setup.

The vendor above is Digital Tigers in Atlanta, Georgia.

Another OEM we have worked with many times is Colfax-International Inc. in Sunnyvale California.

Our advice is to start slowly with an entry-level setup to start with and then once you have gotten familiar with the kit you will soon be able to decide on how to expand or even completely upgrade to your personal specifications.

38 Brief Philosophical Digression

Humankind's understanding of the world is still basic and that includes our mathematical tools. We have been working on the assumption for centuries that things are sort of linear whereas the truth is that the linear state is a great exception to the rule. Our toolkit until quite recently (18th century slow start) was full of linear tools. Nonlinearity took center stage around the middle of the last century. Professor Mandelbrot showed that there is virtually no such thing as 'linear' in nature, showed us how to quantify 'roughness' and dissed the misuse of mathematics in financial calculation using simplifying assumptions (which they needed to make them work at all) giving spurious results and used often because they are easy to calculate.

Our feelings are that the current tool chest, as extensive as it might be, is a child's toy toolset when we set about using it on the complexities of real life price series. We cannot stress enough the enormous complexity and multidimensionality of the markets. Sometimes Edward and Jane feel as if someone just handed them a teaspoon, pointed at this good sized mountain, and said 'move it!'

Nevertheless, if we persevere, and keep turning up, the mountain may just turn out to be made of paper . . .

The markets may be extraordinarily bountiful if one has the passion, the intellect and the tenacity to stay the course

We wish you happy and profitable trading – may all your dreams come true . . .

Jane and Edward

23
Brief Philosophical Digression

39
Information Sources

WEBSITES, JOURNALS, BLOGS

http://www.google.finance.com
http://www.tradethenews.com
http://www.marketwatch.com
http://www.finance.yahoo
http://www.trade2win.com
http://seekingalpha.com/
http://www.fxstreet.com
http://www.tradergav.com/blogs-i-read/
http://www.johnpiperstrading.com/
http://www.santafe.edu

Charting sites, various

arXiv.org e-Print Archive
Econophysics, various
Afraid to Trade Blog
Technical Analysis of Stocks and Commodities (USA)
Futures Magazine (USA)
Active Trader (USA)
SFO Magazine (USA)
Traders Journal (Asia)
Technical Analyst Magazine (UK)
Journal of Trading
Journal of Investing
Traders' Magazine (Germany)
Action Future Magazine (France)
Captain Currency
Chairman Maoxian

Chrisperruna
Downtown trader
Dr. Van Tharp's Blog
HectorTrader
High Probability Trading
Kevin's Market Blog
Lord Tedders Futures & Forex Trading
Make Me Some Money
MarketClub Tradersblog
Misstrade
Move the markets
Neural Market Trends
One Bad Trade
OONR7
Phileo's Picture Windows
Pinoy TRADER
Scout forex
Simply Options Trading
Stockbee
Stocktickr Blog
The 3500
The Chart Strategist
The Kirk Report
The Lonely Trader
The Technical trader.net
Timothy Sykes
Toni Hansen's Blog
Trade For Gain
Trader Eyal
Trader ZBS
TraderMike
Wall St. Warrior

We find that following the fortunes of our selected stocks is both useful and entertaining. On the net we would recommend Bloomberg for news coverage, CNBC wherever available with USA content.

The Wall Street Journal on the net – it's a bit of navigation so we will give you the driving instructions: (www.online.wsj/market data scroll down to 'US Stocks, see all US stocks data' then to /closing market statistics/closing quote tables/NYSE stocks and NASDAQ GLOBAL MARKET. Phew, hope you find it.

The 'home' sites NASDAQ, NASDAQ TRADER, NYSE should be bookmarked.

The *Financial Times* is good. And also is a fixture with us, as is *Barron's* weekly if you can spare the weekend to read it.

A subscription to the magazine 'Technical Analysis of Stock and Commodities' available by subscription from www.traders.com is a good way to keep in touch with the latest thoughts, ideas, products and books. It is the basic bible of the individual trader. In the USA some of the larger bookstores stock it.

Another useful and very professional algo oriented journal is *Automated Trader* (which originates in the UK). It is available on subscription only, from the company's website www.automatedtrader.net. The online site also keeps tabs on developments in the algo space.

There are innumerable blogs on trading, some more useful than others. You should scan the list occasionally to see what the other traders are thinking, saying and doing. Besides keeping you informed this may spark creative ideas, the field is very fluid and new titles appear as older ones vanish in constant churn.

APPENDICES

Appendix A 'The List' of Algo Users and Providers

The list which follows comprises outline descriptions of some of the main algorithmic products used and offered by a selection of the major Tier 1 players. We have compiled this list to give the individual trader a perspective view of the current (mid-2010) algorithmic trading space.

By providing the individual trader with this perspective snapshot view of who the big guns are and where the big guns are currently aimed and, finally, a guess where they are heading, this mini overview of the algorithmic trading terrain should provide the requisite orientation for a better understanding of the evolution of this technology.

When browsing through the list there is also the chance of sparking an idea for an ALPHA ALGO design or strategy which the reader might just be able to develop into the next 'killer algo.'

We must stress that this area is growing and changing extremely quickly and the list that follows is only a snapshot of a rapidly evolving technology.

The market structure is also constantly changing with considerable systemic fragmentation taking place. A major change is the development of the so-called 'Dark Pools.'

These have now become a commonplace feature of the fragmentation taking place in the markets. The Dark Pools are anonymous collections of unit liquidity which give institutional investors the possibility to trade a large number of shares without revealing themselves to the lit market.

The authors would like give their profuse thanks to the companies in the list which follows for the use of their website information and acknowledge all copyrights and registered trademarks.

INSTINET

This is one of the very first actual ECNs. Originally use was restricted to the big block traders but metamorphosed into a full-scale matching network.

The acronym 'ECN' was coined by the SEC and stands for 'electronic communications network.' This was initially intended as a strictly computer operated matching system.

In the late 1990s the ECN proliferated. 'ISLAND' became the day trader's best friend with a spate of others following. It was purchased by Instinet. Many of the others were merged into NASA and NYSE.

Instinet's 'Experts' algorithmic trading suite provides a wide range of tactical, strategic and benchmarking solutions for US, Canadian, European and Asia-Pacific equity markets.

Tactical Strategies

BlockPeg®

- Allow access to liquidity above specified quantity threshold while minimizing impact on the spread.

Cobra

- Aggressively seek hidden and displayed liquidity across Exchanges and lit and Dark Pools using randomization to minimize information leakage.

Impact Minimization

- Passively minimize impact and maximize anonymity by placing orders based on volume profile and current trading activity.

Market Rules HK Access, Japan Short, Taiwan Access

- Automate adherence to local market rules.

Nighthawk®

- Intelligently integrate dark liquidity sources into one destination while taking advantage of embedded anti-gaming and footprint minimization logic.

Pairs Trading, Pairs, Contingent Pairs

- Correlate a pair of securities to remain cash-neutral or share balanced while executing on predetermined spread or ratio.

Work

- Trade an order in as natural a manner as possible dynamically adjusting tactics based on market conditions and minimizing footprint and signaling.

Benchmark Strategies

Auction Target Close Auction

- Participate in opening/closing auctions.

Dynamic Participation DynaPart, Sidewinder

- Dynamically adjust volume participation rates in response to relative price movements.

Implementation Shortfall IS Wizard

- Minimize implementation shortfall at stock level by optimizing balance between expected impact and opportunity cost.

Portfolio IS

- Minimize implementation shortfall at portfolio level by optimizing balance between expected impact and opportunity cost.

TWAP Asia TWAP Asia Tline

- Distribute trades evenly across time while reducing impact and maximizing anonymity through sophisticated execution techniques.

Volume Participation, % of Volume

- Target a user-specified percentage of volume over the specified trading horizon, with user parameters to determine adherence to volume profile.

VWAP Asia VWAP Asia Vline

- Trade against the volume-weighted average price over the specified trading horizon while reducing impact through sophisticated execution techniques.

CREDIT SUISSE

AES is Credit Suisse's sophisticated suite of algorithmic trading strategies, tools and analytics for global securities trading with over a thousand Credit Suisse institutional clients.

CrossFinder+ uses historical and real-time trade information, and determines how to dynamically provide and remove liquidity among fragmented dark crossing venues with objectives to maximize liquidity and minimize market impact. All other AES tactics have integrated CrossFinder+ technology into their decision making formula.

One strategy aimed at minimizing this impact is SNIPER – an aggressive and opportunistic liquidity-seeking algorithm developed by Credit Suisse to pick off liquidity as it becomes available at a target price. Usage of SNIPER has more than doubled during the last 18 months, reflecting the desire of many investors to achieve execution while markets have been volatile.

The AES® suite of algorithms also includes traditional algorithmic strategies that seek to divide trading volumes up over time and strategies that seek to trade at the Volume Weighted Average Price of a stock. Additionally, AES® offers strategies that seek to minimize implementation shortfall – or the difference between the price at which a client decides to trade and the actual execution cost – such as INLINE and other liquidity-seeking strategies like GUERRILLA.

FIDESSA

May 2008: Fidessa Group Plc (LSE:FDSA) announced the launch of BlueBox Execution Strategies, a suite of algorithms designed to provide traders with short-term trading tactics that combine access to Dark Pools along with conventional markets. The Fidessa trading platform in the US, BlueBox Execution Strategies can access and smart route to displayed and/or Dark Pools of liquidity, simultaneously or in sequence, to assist traders in optimally working orders with varying trading objectives.

SPOTLIGHT, a Dark Pool and visible market aggregation algorithm, can simultaneously connect to and access displayed and non-displayed liquidity. The customer's order is split across the available Dark Pools and, based on execution feedback, the algorithm will rebalance order quantity across those pools with active liquidity.

The remaining BlueBox Execution Strategies include: POUNCE, which includes a technical pricing model to take stealth-like advantage of market opportunities; LADDER, which allows users to peg multiple price points within the market; and MIDMAX, which is designed to achieve price improvement by exploiting passive and aggressive trading strategies.

In addition to this launch, the existing suite of standard BlueBox algorithms, designed to achieve certain market benchmarks, has been expanded with the addition of Dark Arrival. This algorithm has been developed specifically to control access to Dark Pools of liquidity. Dark Arrival will route an order quantity sequentially across preferred Dark Pools and will then follow the Arrival Price trading model to achieve optimal results. Existing BlueBox benchmark strategies include out-of-the-box algorithms for VWAP, TWAP, Arrival Price, Percentage of Volume, Corporate Buy Back, Scaler and Exit Price.

Some data about the Fidessa group as an example of the coverage typical of a major Tier 1 company:

Fidessa serves over 22 000 users at around 520 clients around the world and serves over 85% of global, Tier 1 equity brokers. Fidessa's global connectivity network provides links to around 255 brokers, 1500 Buy sides and 92 Exchanges.

DEUTSCHE

Welcome to Autobahn® Equity
Deutsche Bank Autobahn® Equity Releases the Stealth Strategy Algorithm

February 2009: Deutsche Bank's Autobahn® Equity business, within its Global Markets division, announced the global availability of its 'Stealth' strategy algorithm.

The Stealth strategy is liquidity-seeking, and is designed to minimize the traders' market impact by trading only as opportunities arise and minimizing information leakage.

The Stealth algorithm was designed for rapidly changing markets characterized by wider spreads, faster quoting, and smaller top-of-book liquidity coupled with the enormous growth in off-exchange liquidity.

All Deutsche Bank's algorithms, including Stealth, are fully integrated with Deutsche Bank's ultra-fast smart order router, which adapts the execution of trades to all available dark and displayed venues.

BARCLAYS

Algorithmic Trading Strategies

VWAP – Volume Weighted Average Price

- Distributes large orders in line with the expected market volume in order to attain the VWAP benchmark and minimize market impact. The level of aggression can be altered to increase the opportunity of capturing the spread.

TWAP—Time Weighted Average Price

- Distributes your order into equal slices and executes over specified time interval to minimize market impact. The level of aggression can be altered to increase the opportunity of capturing the spread.

Participate

- Target a percentage of market volume. The strategy dynamically adapts to volume conditions throughout the life of the order.

Reference/Arrival Price

- You can define a participation rate when an order is 'in the money' and when an order is 'out of the money' compared to a reference price. When market prices are advantageous the strategy will speed up participation (or vice versa). If no reference price is submitted the mid price upon arrival will be used.

Slice

- You can slice up an order into many 'child' orders in one go. Slice increases trader efficiency by sending simultaneous orders to work within exchange thresholds.

Strategy/Pairs Trading

- This algorithm takes two or more contingent orders and executes them according to a predefined relationship between them. Often this is used to trade a spread between a pair of instruments. However this strategy can be used to trade any relationship between any number of securities (or legs).
- The strategy can operate in both 'basic' and 'speculative' mode.
  - **Basic Mode:** The strategy observes the prices and quantities trading on the market and when the correct trading conditions are met, orders are then sent to the market.
  - **Speculative Mode:** Selected legs of the trade are 'shown to market' at a level that would satisfy the relationship. If any of these legs are filled, the other legs are then executed.
  - By being in the market, speculative mode increases the possibility of trading at desirable levels. Slick correction features are built in to compensate for any previous price slippage that may be encountered during the execution of the algorithm.

GOLDMAN SACHS

Through the REDIPlus Execution Management System, or via FIX connectivity Goldman Sachs provides full access to its multi-product algorithms, integrated portfolio and spread trading functionality, a suite of pre- and post-trade analytics, as well as clearing and prime brokerage services.

Algorithmic Trading

Goldman Sachs Algorithmic Trading (GSAT) offers a global, multi-asset suite of algorithms that includes equities, futures, synthetics and options. Every algorithm seeks optimal execution based on user-defined parameters and accesses a number of displayed and non-displayed sources via our SIGMA smart router. Additionally, clients have the opportunity to cross orders with non-displayed liquidity in our SIGMA X crossing network. You can connect to GSAT Goldman Sachs algorithms in a variety of ways, including REDIPlus, Bloomberg, Order Management Systems or FIX.

Start-of-day position feeds ensure that positions are available in the REDIPlus EMS. All trades executed via REDIPlus are seamlessly delivered to accounts. In addition, REDIPlus has electronic access to Goldman, Sachs & Co. stock lending desk.

VWAP – Volume-Weighted Average Price

- Minimize shortfall relative to a Volume-Weighted Average Price (VWAP) benchmark.

PARTICIPATE/POV – Volume Participation (In Line)

- Work an order with the volume on exchange, with a limit placed on shares traded relative to total volume traded on the order book.

4-CAST – Optimal Execution

- Optimize the trade-off between market impact and opportunity cost with the objective of minimizing the implementation shortfall.

PICCOLO/MIDAS – Small Order Spread Capture

- Improve execution price on marketable orders by capturing the spread on the inside quote.

AUCTION – Market/Limit on Close

- Automate delivery of orders in opening, intraday and closing auctions for European markets.

GSAT NAVIGATOR – Algorithm of Algorithms

- Select the most appropriate algorithmic strategy for the order parameters entered by the trader.

ACTIVE AT PRICE – Target Price

- Work order aggressively once stock trades better than target price.

TWAP – Time Weighted Average Price

- Minimize shortfall relative to a Time-Weighted Average Price (TWAP) benchmark.

AUTO SHORT-SELL – Uptick Price

- Help traders execute short-sell orders by automatically offering stock at the prevailing uptick price.

SHORT-SALE

- Optimize the execution of short sale orders that must adhere to the uptick rules.

DYNAMIC SCALING

- Advanced form of the percentage of volume algorithm where the target participation rate dynamically adjusts to the prevalent price level with respect to a benchmark selected by the user.

SMALL CAP

- Execute stocks for which liquidity is the main concern (seeks hidden liquidity, places orders using smart order types, monitors market impact on past trades and available opportunities).

CITIGROUP

The cost of creating and then keeping up to date new proprietary algos is relatively high for new entrants as it is a fairly drastic technology paradigm shift. Those companies who have not developed their own research teams have resorted to buying competing companies.

Citigroup bought Lava Trading and OnTrade and is said to have paid nearly \$700 million for Automated Trading Desk (NASDAQ ADSK) a company which is estimated to have 6% of North American trading volume (guess = 200 million shares per day). Analysts believe that it will take Citi several years to integrate these companies and to catch up with the current market leaders.

In late March 2010 Citigroup is introducing a new set of algorithmic strategies, as well as technology focused on speed and geared to an electronic marketplace with dozens of venues. Citigroup's almost 1000 clients will make the switch to the new trading products by May 1, when the old system is retired. The company targets an increase in market share from its current 14% to 20%.

We have not detailed the product offer as the information is still too fluid at the time of writing.

NEONET

Stockholm, September 2009 – Neonet enhanced its dark offering by launching NEONET DARK, a dark liquidity consolidating algorithm. The dark algorithm minimizes market impact by finding dark liquidity in available Dark Pools through Neonet's liquidity network. Neonet Dark systematically sweeps the Dark Pools for non-displayed liquidity.

'The development of Europe's market structure has driven the demand for a new generation of algorithms that can provide access to non-displayed liquidity. Neonet Dark gives our customers the ability to seek liquidity in multiple dark pools by sending one consolidated order. Access to all significant pools of liquidity is essential for our clients and therefore we continue to develop and provide solutions that facilitate access to both lit and dark liquidity,' says Simon Nathanson, CEO and President of Neonet.

Neonet Dark is the most recent addition to Neonet's comprehensive suite of solutions for electronic execution.

UBS

UBS is one of the early major players in the algorithmic and quantitative space from circa 1990. The then chairman of UBS, Marcel Ospel, took a great personal interest in the work involving chaos theory and financial prediction of Professors Farmer and Packard and their company located in Santa Fe, New Mexico – The Prediction Company. (In the book of the same name John Bass chronicles the startup of the company, please see our Bibliography.)

This interest resulted in an exclusive contract for financial work with UBS and over the course of time The Prediction Company was eventually bought by UBS outright. See Table A.1.

FIDELITY CAPITAL MARKETS

FCM's Proprietary Algorithmic Strategies include:

Liquidity Seeking Algorithms

- FCM's Liquidity Seeking Algorithms aim to source liquidity leveraging DarkSweep® and an opportunistic model around price and size. As liquidity is sourced, the algorithms rebalance the allocation in order to capture all available liquidity. The algorithms are designed to balance the anonymity of a trader's intentions and getting an order completed when favorable liquidity or price levels appear. None of the algorithms will post your order out loud.

AdrenalineSM

- AdrenalineSM will survey quoted venues in an attempt to identify significant levels of liquidity on the bid or offer while working a piece in DarkSweep®. Once triggered by the bid/offer size exceeding the symbol-specific characteristics, AdrenalineSM will access liquidity based on Aggressiveness. Aggressiveness determines how sensitive AdrenalineSM is and how much of the quoted liquidity to attempt to take.

RecoilSM

- RecoilSM looks for local price momentum while searching for liquidity in DarkSweep®. Using symbol-specific characteristics/tics, RecoilSM determines when favorable price points appear. When triggered, the algorithm begins intelligently interacting with liquidity from both dark and quoted venues.

T-HawkSM

- T-HawkSM patrols for advantageous liquidity and price levels while representing your order in DarkSweep®. The algorithm sources liquidity from dark and quoted markets when the opportunity presents itself.

FCM's Traditional Algorithmic Strategies include:

Table A.1 UBS algorithms

UBS Strategy	Description	Key Control	Region
TapNow	A very aggressive liquidity-seeking strategy which aggregates liquidity across displayed markets and dark pools.	Always Aggressive	Global
Tap	Simultaneously seeks liquidity in both displayed and non-displayed markets, based on your Urgency setting and parameters.	Urgency	Global
Volume Inline	Seeks a target participation rate with respect to overall market volume.	Volume, Price Limit	Global
Price Inline	Trades with sensitivity to a trader defined benchmark. Modifies participation based on price movement relative to the benchmark, but still targets completion.	Urgency, Reference Price	Global
Implementation Shortfall	Executes along a front loaded trajectory while optimizing the trade-off between market impact and risk in relation to the selected Urgency.	Urgency	Global
VWAP	Targets liquidity demand with a stock's expected volume pattern over your specified period. Orders are randomized to minimize market visibility.	End Time	Global
TWAP	Targets an execution while keeping pace with the time, resulting in a linear execution evenly spaced over the entire duration. Orders are randomized to minimize market visibility.	End Time	Global
Pairs	Executes two stocks simultaneously, while monitoring prices for each and trading only when the target spread can be achieved. Does not participate in Opening or Closing auctions.	Price Limit	US, Europe
Perimeter	Enables trading outside of regular US market hours: pre-open, post-close, in auctions and imbalances. Available from 7:30am – 5:30pm.	Session Type	US
At Open	Places orders in the opening auction. Can continue after the auction if Discretion parameter applied.	Price Limit	Europe, APAC
At Close	Places order in the closing auction. Can start early if Discretion parameter applied.	Price Limit	Global
Float	Places order on the bid, midpoint or offer and then floats with the market movement.	Style	Europe, APAC
Customized Algorithms	All strategies may be customized to modify their behavior based on your preferred trading style. We can also build custom algorithms to suit your specific needs.		Global
UBS Tactic	Description	Key Control	Region
DMA	Direct Market Access. Where multiple liquidity pools exist, UBS applies smart order routing, allowing you to execute orders directly in the market place with high capacity links to all major market centers.	Price Limit	Global

Hidden	Holds order off exchange until desired price is available.	Price Limit	Europe, APAC
Stop Loss	Buys or sells once stock reaches a specific price.	Stop Price	APAC
Sweep Display	Aggressively sweeps displayed liquidity at the inside of the market until the order is filled or is no longer marketable (limit display price reached or no fills received).	Display Size	US
Sweep No Display	Aggressively and continuously sweeps displayed liquidity – does not timeout and does not post. Orders will sit in the smart router until filled, cancelled, or the EOD is reached.		US
Deep Sweep	A Deep Sweep Limit order uses ISOs to go to an immediate limit price, while a Deep Sweep Market order uses a 'SweepLevel' parameter to determine how far down to execute under the NBBO with each pass. A Deep Sweep market order will iterate until the order is filled or 1 second has elapsed.	Sweep Level	US
Sweep IOC	A single sweep across the displayed market, followed by an immediate cancel on the remainder.		US
Auction Only	Participates in the opening and/or closing auction.	Include Open/ Close	US
UBS Options Strategy			
Options Delta Adjust	Executes a limit order based on a function of delta related to the reference price.	Reference Price	US
Options TWAP	Releases orders into the market at an even pace over a specified time period, with a parameter for delta to manage deviation.	End Time	US
Options Trigger	Places contingent orders based on price movements around your reference price.	Reference Price Trigger	US
Options Scale Trigger	Places contingent orders based on price movements around your minimum and maximum reference prices, becoming more aggressive as the price increases.	Max/Min Reference Price Triggers	US
UBS Futures Strategy			
Futures VWAP	Seeks a volume-weighted average series of executions for your order.	End Time	US
Futures TWAP	Releases orders into the market at an even pace over a specified time period.	End Time	US

VWAP – Volume Weighted Average Price

- Ensures that the trade will take place at the average-weighted price of the day or lower (if it is purchased) or higher (if it is a sale). VWAP will fill an order in its entirety.

SNAP – Implementation Shortfall; Arrival Price

- Measures the stock price at the time of the order, throughout the day, and at the time of execution. Using an Arrival Price (midpoint of Bid/Ask at time of arrival) benchmark, SNAP will determine the best execution tactic while working around the benchmark, given a user-specified Urgency level. SNAP will fill an order in its entirety.

TVOL – Target Volume

- Allows you to participate in the market at the rate specified by the trader. The participation rate may be modified at any time by the trader. TVOL will not necessarily fill an order in its entirety.

TWAP – Time Weighted Average Price

- Uses a flatter model, indicating more evenly spaced trades throughout the day. Traders can adjust the model to control the concentration of volume over the time interval. Volume can be pulled forward for momentum orders, or pushed back (for value orders). TWAP will fill an order in its entirety.

TPRC – Target Price

- TPRC will be more aggressive if the market moves in your favor and less aggressive if it moves against. It incorporates real-time market data to adjust the trader's selected price. TPRC will not necessarily fill an order in its entirety.

MOC – Market on Close

- This model is designed to improve upon the actual closing price. It also allows for trading before the close if it judges that the order will incur material market impact. MOC will fill an order in its entirety.

BUYB – Corporate Buyback

- This strategy will attempt to aggressively take stock whenever eligible, in full compliance with 10b-18. The aggressiveness of the strategy may be adjusted by the Urgency level, where Low is the least aggressive strategy, and High is extremely aggressive. BUYB will not necessarily fill an order in its entirety.

FADE

- This passive strategy minimizes the use of market orders throughout the trading interval. However, you can specify a guaranteed fill of the order at your discretion. In this case, if necessary to complete the order, the algorithm will revert to using market orders near the end of the trading interval. FADE will fill an order in its entirety if Guaranteed Fill is checked.

Other Tier 1 Players

In order to round off this list of firepower we would like to mention some of the other players in the 'algorithmic space' whose product offering we have not detailed:

JP MORGAN, PARIBAS, PORTWARE, APAMA, BKNY CONVERGEX, BANK OF AMERICA, MERRILL LYNCH, WACHOVIA, MORGAN STANLEY AND MANY OTHERS.

Appendix B
Our Industry Classification SECTOR Definitions

Automotive Transport

Passenger cars, trucks, long haul vehicles, motorcycles, bicycles.

Consumer Discretionary

White goods, brown goods, household durables, apparel, leisure equipment, sports equipment.

Consumer Staples

Food, beverages, tobacco, cosmetics.

Defense

Airforce, Navy, armoured vehicles, infrastructure equipment, weapons.

Energy

Oil Rig construction and provision, supply of infrastructure drilling equipment
Natural resources exploration and analysis
Rig operation
Refining transportation retailing of oil and gas products
Coal mining and distribution
Solar power.

Financial

Banking retail, banking corporate, investment banking, hedge funds, brokerages, insurance, venture capital.

Healthcare

Medical equipment, healthcare facilities.

Industrial

Manufacture and distribution of capital goods, industrial white and brown goods, construction, electrical equipment, industrial machinery. Provision of commercial services and supplies, printing, office services, employment agencies, couriers.

Information Technology

Internet businesses, database providers, network operators, ISPs.

Materials

Manufacture of chemicals, glass, fiberoptics, mining and refining metals and minerals, steel producers, production of paper, paper products, packaging, plastics, raw materials.

Pharmaceutical

Research, development, marketing pharmaceuticals.

Real Estate

Real estate operation, development.

Semiconductors

R&D manufacture and marketing of integrated circuits, memory, CPUs, and other electronic devices.

Technology

Computer hardware manufacture and distribution, software development and distribution.

Telecom

Provision of communication services, fixed line copper, fiber optic, cellular wireless, cable.

Transport

Airlines, road transport, railways, boats, ships and tankers.

Utilities

Electric, gas, water.

Appendix C
The Stock Watchlist

This list of stocks is our own 'Watchlist' of issues which we monitor on an ongoing basis and have found to be tradable using our ALPHA ALGOS.

There is considerable variability in trading performance and a lot of iterations and empirical work which has to be done all the time to keep up with the evolution of the market. A stock which has been trading exceptionally well may suddenly, and without any apparent reason, stop responding. Parameters have to be checked and updated when necessary.

Some of the market property changes used to be fairly gradual but in recent years these changes (not counting meltdown-type changes) seem to have become more abrupt and of a higher amplitude. These systemic changes plus regulatory changes all impact the performance of algorithms and trading in general.

As may be expected the components of the Watchlist are also a moving target with some stocks being added and some being deleted over the course of time as their characteristics and properties change. The tradability of individual stocks is constantly in a state of flux.

Over time you will find certain stock/algo combinations that work for you especially well. Our advice is to search and collect these and keep trading them and constantly checking on performance and keeping logs of all the metrics. The profit power of a particular combination fluctuates, like all market systems, so do not be too surprised if these swings are quite wide, from minimal profit efficiency to exceptional returns.

Decide on how many stocks you want to follow initially. We would suggest that you select 20 to start with. It is best to copy the 'DATA' for each session to an external hard disk on a daily basis.

EOD metrics should be plotted daily and converted to a series with an occasional much more detailed fine grained analysis intraday. Let the data speak or, rather, whisper its secrets.

Finally it is a good idea to make a weekly summary of which stocks did best and with which algos – keep it in your 'STOCK PROFILING' file. This could serve as a starting point for the following week's trading plan.

The following list is also provided on the CD filename 'WATCHLIST.' It is heavily biased toward the pharmaceutical and oil exploration sectors as we have found those clusters easier to work for new traders at the time of writing. However, keep in mind that this can all change in a heartbeat – *vide* the Gulf of Mexico disaster. In stock selection the key words are 'be informed' and 'be vigilant.'

AAPL	FLR
ABT	GD
ADBE	GE
ADSK	GENZ
AKAM	GILD
ALTR	GLW
AMAT	GOOG
AMD	GS
AM	HAL
GN	HES
AMZN	HUM
ANF	MERK
APOL	NOC
BA	ICE
BBBY	OIH
BIIB	OXY
BIDU	PCAR
BLK	PFE
CAT	RIG
CELG	RTN
CEPH	SBUX
CM	SHLD
E	SNDK
DO	TDW
ESRX	UPS
ESV	WYNN
FCX	YHOO
FDX	YUM

Appendix D
Stock Details Snapshot

IMPORTANT: Please regard this chapter as a 'snapshot' in time. The values change in every trading session. The values in this chapter are for that particular EOD only. We are providing them here to give the individual trader a head start in getting familiar with a small number of stocks.

Being familiar with the ticker symbol and its stock name is a good start to feeling comfortable in the arena. We find it useful to know something about the operations of the company though many practitioners may disagree with our view.

We would suggest that, as a start, the very first time, you fill in current values for Trade Price, Volume, Range, Market Cap, and Share Outstanding in the printed section. When you have connectivity and a data feed set up we would advise completing the update of the tables in the book. We have left titled blanks in the tables which you can fill in once you have your data feed for SHARES/TXN, TRAVERSE (EOD), and SHARPE 150T AVERAGE.

Thereafter, on a routine basis, fill in the EXCEL spreadsheet version on the CD. These are available on many websites such as www.finance.yahoo.com, www.marketwatch.com and www.google.finance.com; your brokerage should also provide most of the real-time data you need.

It is worthwhile to read through the business description of the companies' operations which we have included in the book. We have also, where relevant, included our comments on our experience trading the individual stocks. Obviously all this information dates extremely rapidly over time but at least it provides a starting point structure from which to develop your own analyses.

We have found when using a variety of algorithms that we seem to make better parametric decisions when we are thoroughly familiar and 'involved' with the stock's characteristics and 'personality.'

To a certain extent, the better the 'friendship' with the individual stock the better the return is likely to be.

The decision to include such out-of-date information in print relating to a target moving at warp speed was based on our belief that seeing it first in print and then modifying the printed page has a psychological value of 'ownership' which is then further reinforced by working the same material electronically.

Set up your own database of tick data for a selection of stocks (one file per stock) which you will start trading and update it on every trading day at EOD. This is quite a chore but will be most valuable as it builds up. It provides 'backtest' material and an experimental test bed to try out ideas.

This is best stored on an external hard drive as it will eat up space at quite a rate, depending of course on how many stocks you decide to track. We use an array of 4 × 750GB drives.

Also please refer back to the 'Our Metrics' chapter in the book and also the 'ALGOMETRICS' file on the CD for various analysis ideas and how to calculate SHARES/TXN, TRAVERSE (EOD), and SHARPE 150T. After the print 'initiation' please go to the CD file 'STOCK DETAIL SNAPSHOT TEMPLATES.' The tabs may be populated with the stock symbols you wish to monitor. The file on the CD contains the same headings as those set out in the book.

AAPL

AAPL		**3rd QTR 2009**	
DATE	**30 10 09**	Mkt cap	168.86
Open	196.81	Shares	895.83
Close	188.50	EPS	5.72
MIN	186.07	P/E	32.94
MAX	196.80	BETA	1.62
Range	10.73	INST	72
%RANGE	0.05605		
VOL	25.63M		
TXN			
SHARES/TXN			
TRAVERSE			
SHARPE 150T			

Company Description

Apple Inc. (Apple) designs, manufactures and markets personal computers, mobile communication devices, and portable digital music and video players, and sells a variety of related software, services, peripherals and networking solutions. The company sells its products worldwide through its online stores, its retail stores, its direct sales force, and third-party wholesalers, resellers and value-added resellers. In addition, the company sells a variety of third-party Macintosh (Mac), iPhone and iPod compatible products, including application software, printers, storage devices, speakers, headphones and various other accessories and peripherals through its online and retail stores, and digital content and applications through the iTunes Store. The company sells to consumers, small and mid-sized business (SMB), education, enterprise, government and creative customers.

AAPL – RELATED COMPANIES				
	3rd QTR 2009	Valuation		Profile
	Company name	Price	Mkt Cap	Employees
AAPL	Apple Inc.	188.5	168.86B	34,300
MSFT	Microsoft Corporation	27.73	247.58B	93,000
HPQ	Hewlett-Packard Company	47.46	112.53B	321,000
GOOG	Google Inc.	536.12	169.72B	19,786
DELL	Dell Inc.	14.49	28.34B	76,500
ADBE	Adobe Systems Incorpor...	32.94	17.25B	7,335
T	AT&T Inc.	25.67	151.45B	284,970
PALM	Palm, Inc.	11.61	1.65B	939
INTC	Intel Corporation	19.11	106.98B	80,800
NOK	Nokia Corporation (ADR)	12.61	46.75B	123,347
JAVA	Sun Microsystems, Inc.	8.18	6.16B	29,000
Sector: Technology > Industry: Computer Hardware				

Our Comments

We have found AAPL quite difficult to trade. We experienced many short range changes in volatility, with often not enough price movement/second. Not recommended as a starting stock. Market orders at 1000 shares are filled instantaneously. This is one of the most liquid stocks on the NASDAQ. Use LC algo.

ABT

ABT		**3rd QTR 2009**		
DATE	**13 11 09**	Mkt cap	81,899.80	M
Open	53.40	Shares	1,546.74	M
Close		EPS	3.60	
MIN	52.78	P/E	14.72	
MAX	53.50	BETA	0.18	
Range	0.72	INST	67.00	%
%RANGE	0.014			
VOL	5.96M			
TXN				
SHARES/TXN				
TRAVERSE				
SHARPE 150T				

Company Description

Abbott Laboratories is engaged in the discovery, development, manufacture and sale of a range of healthcare products. It has four segments: Pharmaceutical Products, Nutritional Products, Diagnostic Products and Vascular Products. Pharmaceutical Products include a line of adult and pediatric pharmaceuticals manufactured, marketed and sold worldwide. Diagnostic Products include diagnostic systems and tests manufactured, marketed and sold worldwide to blood banks, hospitals and commercial laboratories. Nutritional Products include a line of pediatric and adult nutritional products manufactured, marketed and sold worldwide. Vascular Products include a line of coronary, endovascular and vessel closure devices manufactured, marketed and sold worldwide. In January 2009, it acquired Ibis Biosciences, Inc. In February 2009, it acquired Advanced Medical Optics. In October 2009, it acquired Visiogen.

Our Comments

Another pharma with a nutritional twist responds well to LC.

ADBE

ADBE		**3rd QTR 2009**		
DATE	**03 11 09**	Mkt cap	17.31	B
Open	32.60	Shares	523.76	M
Close	33.00	EPS	1.25	
MIN	32.37	P/E	26.5	
MAX	32.98	BETA	1.82	
Range	0.61	INST	89	%
%RANGE	0.019			
VOL	3.98			
TXN				
SHARES/TXN				
TRAVERSE				
SHARPE 150T				

Company Description

Adobe Systems Incorporated (Adobe) is a diversified software company. The company offers a line of creative, business and mobile software and services used by creative professionals, knowledge workers, consumers, original equipment manufacturer (OEM) partners, developers and enterprises for creating, managing, delivering and engaging with content and experiences across multiple operating systems, devices and media. It distributes its products through a network of distributors, value-added resellers (VARs), systems integrators, independent software vendors (ISVs) and OEMs, direct to end users and through its own website at www.adobe.com. It also licenses its technology to hardware manufacturers, software developers and service

ADBE – RELATED COMPANIES				
3rd QTR 2009		Valuation		Profile
	Company name	Price	Mkt Cap	Employees
ADBE	Adobe Systems Incorpor...	32.65	17.09B	7,335
MSFT	Microsoft Corporation	27.5	244.26B	93,000
CRM	salesforce.com, inc.	58.33	7.24B	3,653
AAPL	Apple Inc.	188.11	169.45B	34,300
GOOG	Google Inc.	535.15	169.38B	19,786
SNE	Sony Corporation (ADR)	28.69	28.80B	170,800
CREL	Corel Corporation (USA)	3.7	95.85M	1,040
ADSK	Autodesk, Inc.	25.1	5.77B	7,800
IBM	Intl. Business Machine...	120.4	157.78B	398,455
INTC	Intel Corporation	18.42	103.17B	80,800
HPQ	Hewlett-Packard Company	47.65	112.91B	321,000
Sector: Technology > Industry: Computer Software				

providers, and offers integrated software solutions to businesses of all sizes. Adobe has operations in the Americas, Europe, Middle East and Africa (EMEA) and Asia. In September 2008, it acquired YaWah ApS, a dynamic imaging software provider based in Denmark.

Our Comments

We have found ADBE comfortable to trade and often responding well to ALPHA-1. No great surprises but steady. Often a good idea to set a 'money stop' around $175 depending on volatility swings.

ADSK

ADSK		**3rd QTR 2009**		
DATE	**03 11 09**	Mkt cap	5.8	B
Open	25.10	Shares	229.73	M
Close	25.23	EPS	−0.11	
MIN	24.73	P/E		
MAX	25.25	BETA	2.25	
Range	0.52	INST	92	%
%RANGE	0.021			
VOL	2.8M			
TXN				
SHARES/TXN				
TRAVERSE				
SHARPE 150T				

ADSK – RELATED COMPANIES				
	3rd Quarter 2009	Valuation		Profile
	Company name	Price	Mkt Cap	Employees
PMTC	Parametric Technology	15.2	1.77B	5,087
ORCL	Oracle Corporation	20.89	104.73B	84,639
MSFT	Microsoft Corporation	27.56	244.71B	93,000
IBM	Intl. Business Machine...	121.16	158.83B	398,455
DASTY	Dassault Systemes SA	56.6	6.70B	7,459
CREL	Corel Corporation (USA)	3.71	96.11M	1,040
CIMT	Cimatron Ltd.	1.16	10.73M	298
ANSS	ANSYS, Inc.	41.62	3.67B	1,600
ADSK	Autodesk, Inc.	25.23	5.80B	7,800
ADBE	Adobe Systems Incorpor...	33	17.28B	7,335
AAPL	Apple Inc.	188.75	170.00B	34,300
Sector: Technology > Industry: Software and Programming				

Company Description

Autodesk, Inc. is a design software and services company. The company offers products and solutions to customers in the architectural, engineering and construction, manufacturing, geospatial mapping, and digital media and entertainment markets. The company's two-dimension (2D) horizontal design solutions include AutoCAD and AutoCAD LT. In addition, it offers a range of 2D and 3D discipline-specific design and documentation tools. The company is organized into four reportable operating segments: Platform Solutions and Emerging Business and Other (PSEB), which accounted for 44% of its net revenue during the year ended January 31, 2009 (fiscal 2009), Architecture, Engineering and Construction (AEC), which accounted for 23% of its net revenue in fiscal 2009, Manufacturing Solutions (MSD), which accounted for 21% of its net revenue in fiscal 2009, and Media and Entertainment (M&E), which accounted for 11% of its net revenue in fiscal 2009. It serves customers through distributors in the Americas, Europe, the Middle East, Africa, and Asia Pacific. The company was founded in 1982 and is headquartered in San Rafael, California.

Our Comments

Erratic but occasionally can be very well behaved and works OK with ALPHA-1 though we have found that it does not provide as many triggers as some of our other stocks on the Watchlist.

AKAM

AKAM		**3rd QTR 2009**		
DATE	**04 11 09**	Mkt cap	3.94	B
Open	22.80	Shares	174.4	M
Close	22.85	EPS	0.79	
MIN	22.63	P/E	29.1	
MAX	23.24	BETA	0.91	
Range	0.61	INST	83	%
%RANGE	0.027			
CCVOL	4.51M			
TXN				
SHARES/TXN				
TRAVERSE				
SHARPE 150T				

Company Description

Akamai Technologies, Inc. (Akamai) provides services for accelerating and improving the delivery of content and applications over the Internet, from live and on-demand streaming videos to conventional content on websites, to tools that help people transact business. The company's solutions are designed to help businesses, government agencies and other enterprises. It offers services and solutions for digital media and software distribution and storage, content and application delivery, application performance services and other specialized Internet-based offerings. In November 2008, Akamai completed its acquisition of aCerno, Inc.

AKAM – RELATED COMPANIES

3rd Quarter 2009		Valuation		Profile
	Company name	Price	Mkt Cap	Employees
AKAM	Akamai Technologies, Inc.	22.85	3.94B	1,682
GOOG	Google Inc.	548.65	173.69B	19,786
HPQ	Hewlett-Packard Company	48.85	115.83B	321,000
IBM	Intl. Business Machine...	123.10	161.70B	398,455
INAP	InterNAP Network Services	3.18	161.31M	430
LLNW	Limelight Networks, Inc.	3.55	299.82M	301
MSFT	Microsoft Corporation	28.41	252.26B	93,000
RAX	RAXRackspace Hosting, ...	18.07	2.19B	2,648
SDXC	Switch & Data Faciliti...	17.77	614.34M	339
SVVS	SAVVIS, Inc.	14.62	793.87M	2,228
VRSN	Verisign, Inc.	24.00	4.63B	3,297
	Technology	Computer services		

Our Comments

This stock appears to overreact to news in the past. ALPHA-1 has been very successful with this stock. The moves are usually quite abrupt, both ways. As always make sure you put in the capital protection stop as soon as you have put on the trade. We also use LC oscillator.

ALTR

ALTR		3rd QTR 2009		
DATE	**06 11 09**	Mkt cap	5.84	B
Open	19.90	Shares	295.37	M
Close	19.77	EPS	0.78	
MIN	19.68	P/E	25.33	
MAX	20.03	BETA	0.99	
Range	0.35	INST	95	%
%RANGE	0.018			
VOL	4.62M			
TXN				
SHARES/TXN				
TRAVERSE				
SHARPE 150T				

Company Description

Altera Corporation (Altera) is a supplier of programmable semiconductors and related products. Altera designs, manufactures and markets programmable logic devices (PLDs), hardcopy application-specific integrated circuits (ASICs), pre-defined design building blocks, known as intellectual property (IP) cores, and associated development tools. PLDs, which consist of field-programmable gate arrays (FPGAs) and complex programmable logic devices (CPLDs), are standard semiconductor integrated circuits, or chips, that its customers program to perform desired logic functions in their electronic systems. Its HardCopy enables its customer to transition from high-density FPGAs to non-programmable implementations for volume production. IP cores can be licensed by customers to add standard functions to their PLD designs.

Our Comments

In the earlier part of this century this was a very good trading stock. In the recent past not so. Let us see if it recovers to ALPHA-1 in the future.

AMAT

AMAT		**3rd QTR 2009**		
DATE	**07 11 09**	Mkt cap	16,538	B
Close	12.40	Shares	1334	M
		EPS	0.16	
MIN	12.25	P/E		
MAX	12.58	BETA	1.05	
Range	0.33	INST	81	%
%RANGE	0.027			
VOL	13.43M			
TXN				
SHARES/TXN				
TRAVERSE				
SHARPE 150T				

Company Description

Applied Materials, Inc. (Applied), incorporated in 1967, provides nanomanufacturing technology solutions for the global semiconductor, flat panel display, solar and related industries, with a portfolio of equipment, service and software products. The company's customers include manufacturers of semiconductor wafers and chips, flat panel liquid crystal displays (LCDs), solar photovoltaic (PV) cells and modules, and other electronic devices. It operates in four segments: Silicon, Applied Global Services, Display, and Energy and Environmental Solutions. On January 31, 2008, Applied acquired Baccini S.p.A. (Baccini), a supplier of automated metallization and test systems for crystalline silicon (c-Si) solar PV cells.

Our Comments

We have traded this stock for over five years and with moderate success using ALPHA-1 but it can be hard work.

AMD

AMD		3rd QTR 2009		
DATE	**11 11 09**	Mkt cap	3,364.80	M
Open	34.82	Shares	667.6	M
Close		EPS	−3.7	
MIN	4.75	P/E		
MAX	5.05	BETA	12.14	
Range	0.30	INST	45	%
%RANGE	0.061			
VOL	20.95m			
TXN				
SHARES/TXN				
TRAVERSE				
SHARPE 150T				

Company Description

Advanced Micro Devices, Inc. is a global semiconductor company with facilities around the world. The company offers x86 microprocessors, for the commercial and consumer markets, embedded microprocessors for commercial client and consumer markets and chipsets for desktop and notebook personal computers (PCs), professional workstations and servers. It offers graphics, video and multimedia products for desktop and notebook PCs, including home media PCs, professional workstations and servers and technology for game consoles. The company operates in two business segments: computing solutions, which includes chipsets, and graphics. In June 2008, the company launched the Puma platform, which was the code name for the mainstream notebook platform. In January 2009, the company launched the Yukon platform. In March 2008, the company introduced the AMD Phenom X3 8000 series triple-core processors. In March 2009, the company completed the spin-off of its manufacturing operations.

Our Comments

Not for novices as it requires understanding of the chip and cpu market and up to the second awareness of breaking news which will require daily parameterization. LC was best so far.

AMGN

AMGN		3rd QTR 2009		
DATE	**07 11 09**	Mkt cap	55,565	M
Open	54.36	Shares	467.7	M
Close		EPS	4.49	
MIN	53.87	P/E	12.7	
MAX	54.94	BETA	0.53	
Range	1.07	INST	79	%
%RANGE	0.020			
VOL	5.50M			
TXN				
SHARES/TXN				
TRAVERSE				
SHARPE 150T				

Company Description

Amgen Inc. is a biotechnology company that discovers, develops, manufactures and markets human therapeutics based on advances in cellular and molecular biology. It markets human therapeutic products in the areas of supportive cancer care, nephrology and inflammation. Its principal products include Aranesp (darbepoetin alfa), EPOGEN (Epoetin alfa), Neulasta (pegfilgrastim), NEUPOGEN (Filgrastim) and Enbrel (etanercept). Aranesp and EPOGEN stimulate the production of red blood cells to treat anemia and belong to a class of drugs referred to as erythropoiesis-stimulating agents (ESAs). On January 4, 2008, the company completed the acquisition of Dompe Biotec, S.p.A.

Our Comments

One of our favorite pharma stocks. ALPHA-1 over five-session parameterization. When it BUY triggers on a 'breakout,' ignore any SELL triggers (except of course the stop loss) and rely on your experience (otherwise known as 'gut feel') when to sell. It tends to run as traders pile into a trend not to be left behind.

AMZN

AMZN		3rd QTR 2009		
DATE	**08 11 09**	Mkt cap	54,642	M
Open	123.00	Shares	432.98	M
Close		EPS	1.7	
MIN	122.68	P/E	74.02	
MAX	126.98	BETA	1.28	
Range	4.30	INST	69	%
%RANGE	0.034			
VOL	13.23M			
TXN				
SHARES/TXN				
TRAVERSE				
SHARPE 150T				

Company Description

Amazon.com, Inc. (Amazon.com) offers services to consumer customers, seller customers and developer customers. The company serves its consumer customers through its retail websites. It offers programs that enable seller customers to sell their products on the company's websites and their own branded websites. It serves developer customers through Amazon Web Services, which provides access to technology infrastructure that developers can use to enable virtually any type of business. In addition, the company generates revenue through co-branded credit card agreements and other marketing and promotional services, such as online advertising. The company's operations are organized into two principal segments: North America and International. In August 2008, Amazon.com purchased Shelfari, a social network for book lovers. In December 2008, Amazon.com announced the completion of its acquisition of AbeBooks. In November 2009, the company completed the acquisition of Zappos.com, Inc.

Our Comments

Always active and works well with ALPHA-2 We use a five-session lookback.

ANF

ANF		**3rd QTR 2009**		
DATE	**07 11 09**	Mkt cap	3,079	M
Open	34.47	Shares	87.95	M
Close		EPS	0.51	
MIN	34.23	P/E	68.01	
MAX	35.40	BETA	1.43	
Range	1.17	INST	95	%
%RANGE	0.034			
VOL	3.57M			
TXN				
SHARES/TXN				
TRAVERSE				
SHARPE 150T				

Company Description

Abercrombie & Fitch Co. (A&F), through its subsidiaries, is a specialty retailer that operates stores and websites. The Company sells casual sportswear apparel, including knit and woven shirts, graphic t-shirts, fleece, jeans and woven pants, shorts, sweaters, outerwear, personal care products and accessories for men, women and children under the Abercrombie & Fitch, abercrombie, Hollister and RUEHL brands. In addition, the company operates stores and a website offering bras, underwear, personal care products, sleepwear and at-home products for women under the Gilly Hicks brand. As of January 31, 2009, the Company operated 1125 stores in the United States, Canada and the United Kingdom. The operating segments of the company include Abercrombie & Fitch, abercrombie, Hollister, RUEHL and Gilly Hicks.

Our Comments

Used to be very profitable on a daily basis but without any reason we can divine has faded from the scene. By the time you read this it will probably be the 'killer' stock.

Try ALPHA-1 with a longer lookback (say 15).

APOL

APOL		3rd QTR 2009		
DATE	**09 11 09**	Mkt cap	8,436	M
Open	55.00	Shares	154.8	M
Close		EPS	3.74	
MIN	54.29	P/E	14.56	
MAX	55.54	BETA	−0.07	
Range	1.25	INST	85	%
%RANGE	0.023			
VOL	1.88M			
TXN				
SHARES/TXN				
TRAVERSE				
SHARPE 150T				

Company Description

Apollo Group, Inc. (Apollo Group) is a private education provider. The company offers educational programs and services both online and on-campus at the undergraduate, graduate and doctoral levels through its wholly owned subsidiaries, The University of Phoenix, Inc. (University of Phoenix), Western International University, Inc. (Western International University), Institute for Professional Development (IPD), The College for Financial Planning Institutes Corporation (CFFP) and Meritus University, Inc. (Meritus). The company has a joint venture with The Carlyle Group (Carlyle), called Apollo Global, Inc. (Apollo Global), to pursue investments primarily in the international education services industry. During the fiscal year ended August 31, 2009 (fiscal 2009), Apollo Global completed the acquisitions of BPP Holdings plc (BPP) in the United Kingdom, Universidad de Artes, Ciencias y Comunicacion (UNIACC) in Chile and Universidad Latinoamericana (ULA) in Mexico.

Our Comments

Fairly erratic in our experience for such a staid product line. Only had good results when it was thinly traded, a paradox we have not experienced anywhere else. ALPHA-1.

BA

BOEING		**3rd QTR 2009**		
DATE	**11 11 09**	Mkt cap	36,969	M
Open	50.78	Shares	726.6	M
Close		EPS	−0.07	
MIN	50.54	P/E		
MAX	51.42	BETA	1.26	
Range	0.88	INST	59	%
%RANGE	0.017			
VOL	5.57M			
TXN				
SHARES/TXN				
TRAVERSE				
SHARPE 150T				

Company Description

The Boeing Company (Boeing) is involved in the design, development, manufacture, sale and support of commercial jetliners, military aircraft, satellites, missile defense, human space flight, and launch systems and services. The company operates in five principal segments: Commercial Airplanes, Boeing Military Aircraft (BMA), Network and Space Systems (N&SS), Global Services and Support (GS&S) and Boeing Capital Corporation (BCC). BMA, N&SS and GS&S comprise the company's Integrated Defense Systems (IDS) business. The Other segment classification principally includes the activities of Engineering, Operations and Technology (EO&T), an advanced research and development organization focused on technologies, processes and the creation of new products. In December 2008, Boeing announced that it has completed its acquisition of Federated Software Group, whose engineering services and software system help track and distribute equipment and personnel for the United States Department of Defense.

Our Comments

Check on current volatility. Learn the company. Read the politics. Especially check the daily %volatility over ten-session lookback. LC.

BAIDU

BAIDU		**3rd QTR 2009**		
DATE	**07 11 09**	Mkt cap	14,212	M
Open	390.56	Shares	34.69	M
Close		EPS	5.67	
MIN	390.01	P/E	72.31	
MAX	410.54	BETA	1.78	
Range	20.53	INST	48	%
%RANGE	0.051			
VOL	2.54M			
TXN				
SHARES/TXN				
TRAVERSE				
SHARPE 150T				

Company Description

Baidu, Inc. (Baidu) is a Chinese-language Internet search provider. The company conducts its operations in China principally through Baidu Online Network Technology (Beijing) Co. Ltd (Baidu Online), its wholly owned subsidiary in Beijing, China. It also conducts its operations in China through Baidu Netcom Science Technology Co. Ltd. (Baidu Netcom), which holds the licenses and approvals necessary to operate the company's websites and provide online advertising services. In January 2008, the company launched a Japanese search service at www.baidu.jp, run by Baidu Japan. The company's Japanese search services enable users to find relevant information online, including web pages, images, multimedia files and blogs, through links provided on its websites.

Our Comments

We have been trading this stock from the word go with very good results considering that it is thought of as so high risk. ALPHA-2 works well.

BBBY

BBBY		**3rd QTR 2009**		
DATE	**07 11 09**	Mkt cap	9765	M
Open	36.97	Shares	262.69	M
Close		EPS	1.75	
MIN	36.65	P/E	21.27	
MAX	37.35	BETA	1.17	
Range	0.70	INST	95	%
%RANGE	0.019			
VOL	2.58M			
TXN				
SHARES/TXN				
TRAVERSE				
SHARPE 150T				

Company Description

Bed Bath & Beyond Inc., along with its subsidiaries, is a chain of retail stores, operating under the names Bed Bath & Beyond (BBB), Christmas Tree Shops (CTS), Harmon and Harmon Face Values (Harmon) and buybuy BABY. The company sells an assortment of merchandise, principally including domestics merchandise and home furnishings, as well as food, giftware, health and beauty care items, and infant and toddler merchandise. Domestics merchandise includes categories such as bed linens and related items, bath items and kitchen textiles. Home furnishings include categories such as kitchen and tabletop items, fine tabletop, basic housewares and general home furnishings.

Our Comments

Careful during uncertain conditions, recessions and the like. LC.

BIIB

BIIB		**3rd QTR 2009**		
DATE	**13 11 09**	Mkt cap	13,511.36	M
Open	46.33	Shares	289.20	M
Close		EPS	2.99	
MIN	46.31	P/E	15.63	
MAX	47.06	BETA	0.56	
Range	0.75	INST	90.00	%
%RANGE	0.016			
VOL	2.81M			
TXN				
SHARES/TXN				
TRAVERSE				
SHARPE 150T				

Company Description

Biogen Idec Inc. is engaged in the development, manufacturing and commercialization of therapies. The company's products address diseases such as multiple sclerosis, non-Hodgkin's lymphoma (NHL), rheumatoid arthritis (RA), Crohn's disease (CD) and psoriasis. The company has four products: AVONEX (interferon beta-1a), RITUXAN (rituximab), TYSABRI (natalizumab) and FUMADERM (dimethylfumarate and monoethylfumarate salts). AVONEX is used in the treatment of relapsing forms of multiple sclerosis (MS). RITUXAN is one of the top selling oncology therapeutics. In the United States, RITUXAN is approved for NHL. TYSABRI is approved for the treatment of relapsing forms of MS. FUMADERM acts as an immunomudulator. The company also has product candidates, such as BG-12, which is an oral fumarate; ANTI-CD80 monoclonal antibody (MAb)(galiximab); ANTI-CD23 MAb (lumiliximab); Humanized Anti-CD20 MAb (ocrelizumab), Lixivaptan, an oral compound for the potential treatment of hyponatremia, and ADENTRI.

Our Comments

Can be a very tradable stock on occasion. LC.

BLK

BLK		3rd QTR 2009		
DATE	**13 11 09**	Mkt cap	31.30	B
Open	231.31	Shares	134.28	M
Close		EPS	5.01	
MIN	229.56	P/E	46.56	
MAX	233.78	BETA	1.63	
Range	4.22	INST	35	%
%RANGE	0.018			
VOL	126,904			
TXN				
SHARES/TXN				
TRAVERSE				
SHARPE 150T				

Company Description

BlackRock, Inc. (BlackRock) operates as an investment management firm in the United States. As of December 31, 2008, the company had $1.307 trillion of assets under management (AUM). BlackRock provides diversified investment management services to institutional clients and to individual investors through various investment vehicles. Its investment management services primarily consist of the active management of fixed income, cash management and equity client accounts; the management of a number of open-end and closed-end mutual fund families, and other non-United States equivalent retail products serving the institutional and retail markets, and the management of alternative funds developed to serve various customer needs. In addition, BlackRock provides market risk management, financial markets advisory and enterprise investment system services to a broad base of clients.

Our Comments

We have not traded this stock as yet. On first analysis it is ‘interesting.’

CAT

CAT		3rd QTR 2009		
DATE	**07 11 09**	Mkt cap	35,869	M
Open	34.82	Shares	35.86	M
Close		EPS	2.13	
MIN	56.54	P/E	27.03	
MAX	58.20	BETA	1.86	
Range	1.66	INST	62	%
%RANGE	0.029			
VOL	6.16M			
TXN				
SHARES/TXN				
TRAVERSE				
SHARPE 150T				

Company Description

Caterpillar Inc. (Caterpillar) operates in three principal lines of business: Machinery, Engines and Financial Products. Machinery deals with the design, manufacture, marketing and sales of construction, mining and forestry machinery. Engines business deals with the design, manufacture, marketing and sales of engines. Financial Products consist primarily of Caterpillar Financial Services Corporation (Cat Financial), Caterpillar Insurance Holdings, Inc. (Cat Insurance), Caterpillar Power Ventures Corporation (Cat Power Ventures) and their respective subsidiaries. In April 2008, Satyam Computer Services Limited acquired Caterpillar's market research and customer analytics operations. In April 2008, Caterpillar expanded its Global Mining business through the acquisition of Lovat Inc. (Lovat), a global manufacturer of tunnel boring machines used in the construction of metro, railway, road, sewer, water main, penstock, mine access, high voltage cable and telecommunications tunnels.

Our Comments

One of our workhorse producers. Safe (if such a word can safely be used in a book of this genre) and steady. Works perfectly with ALPHA-1 and ALPHA-2. Just needs a little care in setting the trigger parameters. Lookbacks 5 to 10.

CELG

CELG		**3rd QTR 2009**		
DATE	**07 11 09**	Mkt cap	24,389	B
Open	51.55	Shares	459.6	M
Close		EPS	0.79	
MIN	51.55	P/E	67.25	
MAX	53.26	BETA	0.45	
Range	1.71	INST	84	%
%RANGE	0.033			
VOL	3.22M			
TXN				
SHARES/TXN				
TRAVERSE				
SHARPE 150T				

Company Description

Celgene Corporation is an integrated biopharmaceutical company. The company is primarily engaged in the discovery, development and commercialization of therapies designed to treat cancer and immune-inflammatory related diseases. It is involved in research in several scientific areas that may deliver therapies, targeting areas, such as intracellular signaling pathways in cancer and immune cells, immunomodulation in cancer and autoimmunity and placental cell, including stem and progenitor cell, research. The drug and cell therapies it develops are designed to treat life-threatening diseases or chronic debilitating conditions. The Company's commercial stage products include REVLIMID, THALOMID (inclusive of Thalidomide Pharmion), VIDAZA, ALKERAN and FOCALIN. On March 7, 2008, it acquired Pharmion Corporation, a biopharmaceutical company that acquired, developed and commercialized products for the treatment of hematology and oncology patients.

Our Comments

Useful volatility in this pharma stock. Use LC as 'direction finder' not as strict trigger.

CEPH

CEPH		**3rd QTR 2009**		
DATE	**07 11 09**	Mkt cap	4,349.60	M
Open	57.65	Shares	74.66	M
Close		EPS	3.32	
MIN	56.93	P/E	17.57	
MAX	58.26	BETA	0.48	
Range	1.33	INST		%
%RANGE	0.023			
VOL	1.31M			
TXN				
SHARES/TXN				
TRAVERSE				
SHARPE 150T				

Company Description

Cephalon, Inc. is an international biopharmaceutical company engaged in the discovery, development and commercialization of products in three therapeutic areas: central nervous system (CNS), pain and oncology. In addition to conducting an active research and development program, the company markets seven products in the United States and numerous products in countries throughout Europe and the world. Its principal product is PROVIGIL (modafinil) Tablets [C-IV], consisted 51% of its total consolidated net sales, during the year ended December 31, 2008. PROVIGIL is indicated for the treatment of excessive sleepiness associated with narcolepsy, obstructive sleep apnea/hypopnea syndrome (OSA/HS) and shift work sleep disorder (SWSD). Its other principal products are FENTORA (fentanyl buccal tablet) [C-II] and ACTIQ (oral transmucosal fentanyl citrate) [C-II] (including its generic version of ACTIQ (generic OTFC)). In August 2009, the company acquired Arana Therapeutics Limited.

Our Comments

One of our favorite pharma stocks. Very similar to CELG. Use ALPHA – 5 day lookback.

CME

CME		3rd QTR 2009		
DATE	**07 11 09**	Mkt cap	20,367.93	M
Open	303.27	Shares	66.43	M
Close		EPS	10.3	
MIN	299.00	P/E	29.78	
MAX	308.65	BETA	1.17	
Range	9.65	INST	70	%
%RANGE	0.032			
VOL	574,108			
TXN				
SHARES/TXN				
TRAVERSE				
SHARPE 150T				

Company Description

CME Group Inc., formerly Chicago Mercantile Exchange Holdings Inc., offers access to asset classes from a single electronic trading platform and trading floors in Chicago and New York City. The company offers futures and options on futures based on interest rates, equity indexes, foreign exchange, energy, agricultural commodities, metals, and alternative investment products, such as weather and real estate. During the year ended December 31, 2008, the combined volume of CME and CBOT and NYMEX was approximately three billion contracts. As of December 31, 2008, the company's open interest stood at 63 million contracts and its open interest record was 86.1 million contracts set on September 11, 2008. In March 2008, the company acquired Credit Market Analysis Limited. On August 22, 2008, the company completed its merger with NYMEX Holdings, Inc. In April 2009, the company acquired the Carvill Hurricane Index from Carvill America Inc. and renamed it the CME Hurricane Index.

Our Comments

This is probably one of five stocks which rate our 'favorite stock' rosette. If you are careful and disciplined it has been one of our most consistent stocks over the years. The shifts in the realignment of the various Exchanges may change our mind. The high octane pits are soon consigned to history by Leo Melamed (they still call him 'Chairman'), a charming man who, in our opinion, influenced the evolution of the markets with his creativity more than any other single person. (Also writes good science fiction...)

Use ALPHA-2.

DO

DO		3rd QTR 2009		
DATE	**07 11 09**	Mkt cap	13,511.85	M
Open	97.75	Shares	139.01	M
Close		EPS	10.02	
MIN	96.06	P/E	9.7	
MAX	99.29	BETA	0.88	
Range	3.23	INST	95	%
%RANGE	0.033			
VOL	2.36M			
TXN				
SHARES/TXN				
TRAVERSE				
SHARPE 150T				

Company Description

Diamond Offshore Drilling, Inc. (Diamond Offshore) is a global offshore oil and gas drilling contractor. As of December 31, 2008, the company's fleet consisted of 30 semisubmersibles, 14 jack-ups and one drillship. The company offers a range of services worldwide in various markets, including the deep water, harsh environment, conventional semisubmersible and jack-up markets. The company provides offshore drilling services to a customer base that includes independent oil and gas companies and government-owned oil companies. During 2008, the company performed services for 49 different customers with Petroleo Brasileiro S.A., accounting for 13.1% of its annual total revenues.

Our Comments

Same comments as RIG OIH.

ESRX

ESRX		**3rd QTR 2009**		
DATE	**07 11 09**	Mkt cap	23,433.62	M
Open	83.95	Shares	274.72	M
Close		EPS	3.14	
MIN	83.03	P/E	27.17	
MAX	85.30	BETA	0.92	
Range	2.27	INST	89	%
%RANGE	0.027			
VOL	2.65M			
TXN				
SHARES/TXN				
TRAVERSE				
SHARPE 150T				

Company Description

Express Scripts, Inc. is a pharmacy benefit management (PBM) company in North America. The company provides a range of services to its clients, which include health maintenance organizations (HMOs), health insurers, third-party administrators, employers, union-sponsored benefit plans, workers' compensation plans and government health programs. The company operates in two segments: Pharmacy Benefit Management Services (PBM) and Specialty and Ancillary Services (SAAS). Its PBM services include retail network pharmacy management; retail drug card programs; home delivery pharmacy services; benefit design consultation; drug utilization review; drug formulary management programs and compliance and therapy management programs for its clients. On June 30, 2008, the company completed the sale of CuraScript Infusion Pharmacy, Inc. On July 22, 2008, it completed the acquisition of the Pharmacy Services Division of Medical Services Company.

Our Comments

Frequent and short lookback reparameterization ALPHA-3.

ESV

ESV		3rd QTR 2009		
DATE	**07 11 09**	Mkt cap	6,844.60	M
Open	48.07	Shares	141.51	M
Close		EPS	6.41	
MIN	47.60	P/E	7.49	
MAX	49.13	BETA	1.31	
Range	1.53	INST	90	%
%RANGE	0.032			
VOL	1.46M			
TXN				
SHARES/TXN				
TRAVERSE				
SHARPE 150T				

Company Description

ENSCO International Incorporated is an international offshore contract drilling company. The company engages in the drilling of offshore oil and gas wells in domestic and international markets by providing its drilling rigs and crews under contracts with major international, government-owned, and independent oil and gas companies. As of February 17, 2009, the company's offshore rig fleet included 43 jack-up rigs, two ultra-deepwater semisubmersible rigs and one barge rig. In addition, it has six ultra-deepwater semisubmersible rigs under construction. The company is a provider of offshore contract drilling services to the international oil and gas industry. Its operations are concentrated in the geographic regions of Asia Pacific (which includes Asia, the Middle East, Australia and New Zealand), Europe/Africa, and North and South America.

Our Comments

Same as OIH and RIG – we have experimented using multiple trigger levels with ALPHA-1.

FCX

FCX		3rd QTR 2009		
DATE	**07 11 09**	Mkt cap	34,210.80	M
Open	78.58	Shares	430	M
Close		EPS	−33.21	
MIN	78.54	P/E	16.01	
MAX	80.30	BETA	1.83	
Range	1.76	INST	76	%
%RANGE	0.022			
VOL	13.43M			
TXN				
SHARES/TXN				
TRAVERSE				
SHARPE 150T				

Company Description

Freeport-McMoran Copper & Gold Inc. (FCX), through its wholly owned subsidiary, Phelps Dodge Corporation (Phelps Dodge), is a copper, gold and molybdenum mining company. The company's portfolio of assets includes the Grasberg minerals district in Indonesia, which contains the single largest recoverable copper reserve and the single largest gold reserve of any mine; significant mining operations in North and South America, and the Tenke Fungurume development project in the Democratic Republic of Congo (DRC). As of December 31, 2008, consolidated recoverable proven and probable reserves totaled 102.0 billion pounds of copper, 40 million ounces of gold, 2.48 billion pounds of molybdenum, 266.6 million ounces of silver and 0.7 billion pounds of cobalt. Approximately 35 % of its copper reserves were in Indonesia, approximately 31 % were in South America, approximately 28 % were in North America and approximately 6 % were in Africa.

Our Comments

This is a stock we have traded consistently and had good results with ALPHA-1.

Unusual stock which often trades in a range for quite a time then performs spectacular breakouts. ALPHA-2.

FDX

FDX		3rd QTR 2009		
DATE	**07 11 09**	Mkt cap	24,101.43	M
Open	74.45	Shares	312.52	M
Close		EPS	−0.35	
MIN	74.45	P/E		
MAX	77.36	BETA	1.09	
Range	2.91	INST	77	%
%RANGE	0.038			
VOL	2.89M			
TXN				
SHARES/TXN				
TRAVERSE				
SHARPE 150T				

Company Description

FedEx Corporation (FedEx) is a holding company. The company provides a portfolio of transportation, e-commerce and business services through companies that compete collectively, operate independently and manage collaboratively, under the respected FedEx brand. These companies are included in four business segments: FedEx Express, FedEx Ground, FedEx Freight and FedEx Services. Federal Express Corporation (FedEx Express) is the express transportation company, offering time-certain delivery within one to three business days. FedEx Ground provides day-certain service to every business address in the United States and Canada, as well as residential delivery through FedEx Home Delivery. Effective June 1, 2009, Caribbean Transportation Services, Inc. (CTS), a provider of airfreight forwarding services between the United States and Puerto Rico, the Dominican Republic, Costa Rica and the Caribbean Islands, was merged with and into FedEx Express.

Our Comments

Can provide steady profits, use ALPHA-1.

FLR

FLR		3rd QTR 2009		
DATE	**07 11 09**	Mkt cap	8,165.85	M
Open	45.07	Shares	179.86	M
Close		EPS	4.08	
MIN	44.69	P/E	11.12	
MAX	46.07	BETA	1.44	
Range	1.38	INST	87	%
%RANGE	0.030			
VOL	3.12M			
TXN				
SHARES/TXN				
TRAVERSE				
SHARPE 150T				

Company Description

Fluor Corporation (Fluor) is a holding company that, through its subsidiaries, provides engineering, procurement and construction management (EPCM) and project management services. Fluor serves a number of industries worldwide, including oil and gas, chemical and petrochemicals, transportation, mining and metals, power, life sciences and manufacturing. Fluor is also a primary service provider to the United States Federal Government. It performs operations and maintenance activities for major industrial clients, and also operates and maintains their equipment fleet. The company is aligned into five principal operating segments: Oil and Gas, Industrial and Infrastructure, Government, Global Services and Power. Fluor Constructors International, Inc., which is organized and operates separately from its business segments, provides unionized management, construction and management services in the United States and Canada, both independently and as a subcontractor on projects to its segments.

Our Comments

Use ALPHA-1.

GD

GD		3rd QTR 2009		
DATE	**07 11 09**	Mkt cap	25,300.85	M
Open	65.25	Shares	385.8	M
Close		EPS	6.24	
MIN	64.77	P/E	10.51	
MAX	65.82	BETA	1.19	
Range	1.05	INST	79	%
%RANGE	0.016			
VOL	1.68M			
TXN				
SHARES/TXN				
TRAVERSE				
SHARPE 150T				

Company Description

General Dynamics Corporation (General Dynamics) offers a portfolio of products and services in business aviation; combat vehicles, weapons systems and munitions; shipbuilding design and construction, and information systems, technologies and services. The company focuses on delivering products and services to military, federal government, commercial and international customers. General Dynamics operates through four business groups: Aerospace, Combat Systems, Marine Systems, and Information Systems and Technology. On December 19, 2008, the company acquired AxleTech International, a manufacturer and supplier of axles, suspensions, brakes and aftermarket parts for heavy-payload vehicles for a variety of military and commercial customers.

On September 2, 2009, General Dynamics Advanced Information Systems, a business unit of General Dynamics, completed the acquisition of Axsys Technologies, Inc.

Our Comments

Reacts to geopolitical input. Use ALPHA-5 If there is good news support but you must act immediately if you decide there is a trade.

GE

GE		**3rd QTR 2009**		
DATE	**07 11 09**	Mkt cap	163,226.10	M
Open	14.98	Shares	10,647.50	M
Close		EPS	1.12	
MIN	14.83	P/E	13.69	
MAX	15.49	BETA	1.53	
Range	0.66	INST	48	%
%RANGE	0.044			
VOL	165.05M			
TXN				
SHARES/TXN				
TRAVERSE				
SHARPE 150T				

Company Description

General Electric Company (GE) is a diversified technology, media and financial services company. Its products and services include aircraft engines, power generation, water processing, security technology, medical imaging, business and consumer financing, media content and industrial products. As of December 31, 2008, GE operated in five segments: Energy Infrastructure, Technology Infrastructure, NBC Universal, Capital Finance and Consumer & Industrial. In January 2009, the company acquired Interbanca S.p.A., an Italian corporate bank. In April 2008, Oil & Gas completed the acquisition of the Hydril Pressure Controls business from Tenaris. In September 2008, the company announced the sale of its Japanese consumer finance business to Shinsei Bank. During the year ended December 31, 2008, the company acquired Whatman plc; Vital Signs, Inc.; Merrill Lynch Capital, and CitiCapital. In September 2009, the company completed the acquisition of Scanwind.

Our Comments

General for ALPHA-6. Check Vandewalle's topological map.

GENZ

GENZ		3rd QTR 2009		
DATE	**07 11 09**	Mkt cap	14,131.72	M
Open	52.06	Shares	279.31	M
Close		EPS	1.75	
MIN	51.35	P/E	29.81	
MAX	52.50	BETA	0.33	
Range	1.15	INST	91	%
%RANGE	0.022			
VOL	1.39M			
TXN				
SHARES/TXN				
TRAVERSE				
SHARPE 150T				

Company Description

Genzyme Corporation (Genzyme) is a biotechnology company. Genzyme operates in four segments: Genetic Diseases, Cardiometabolic and Renal, Biosurgery and Hematologic Oncology. Genetic Diseases unit develops, manufactures and distributes therapeutic products, with a focus on products to treat patients suffering from genetic diseases and other chronic debilitating diseases, including a family of diseases known as lysosomal storage disorders (LSDs). Cardiometabolic and Renal segment develops, manufactures and distributes products that treat patients suffering from renal diseases, including chronic renal failure and endocrine and cardiovascular diseases. Biosurgery segment develops, manufactures and distributes biotherapeutics and biomaterial-based products, with a focus on products that meet medical needs in the orthopaedics and broader surgical areas. Hematologic Oncology segment develops, manufactures and distributes products for the treatment of cancer.

Our Comments

Occasionally very volatile and thus offers opportunities for ALPHA-1.

GILD

GILD		**3rd QTR 2009**		
DATE	**07 11 09**	Mkt cap	41,831.13	M
Open	45.49	Shares	904.26	M
Close		EPS	2.56	
MIN	45.40	P/E	18.06	
MAX	46.61	BETA	0.46	
Range	1.21	INST	92	%
%RANGE	0.026			
VOL	8.65M			
TXN				
SHARES/TXN				
TRAVERSE				
SHARPE 150T				

Company Description

Gilead Sciences, Inc. (Gilead) is a biopharmaceutical company that discovers, develops, and commercializes therapeutics in areas of unmet medical need. The company has United States and international commercial sales operations, with marketing subsidiaries in Australia, Austria, Canada, France, Germany, Greece, Ireland, Italy, New Zealand, Portugal, Spain, Switzerland, Turkey, the United Kingdom, and the United States. Its commercial team promotes Truvada, Viread, Emtriva, Hepsera, AmBisome, Letairis and Flolan through direct field contact with physicians, hospitals, clinics and other healthcare providers. Gilead's corporate partner, Astellas Pharma, Inc. (Astellas), promotes and sells AmBisome in the United States, Canada, Europe, Australia, and New Zealand. In May 2008, the company acquired Navitas Assets, LLC. In April 2009, the company announced the completion of its acquisition of CV Therapeutics, Inc.

Our Comments

Classic pharma, use ALPHA-1.

GLW

GLW		3rd QTR 2009		
DATE	**07 11 09**	Mkt cap	23,808.23	M
Open	15.02	Shares	1556.09	M
Close		EPS	0.97	
MIN	14.96	P/E	15.8	
MAX	15.48	BETA	1.28	
Range	0.52	INST	77	%
%RANGE	0.034			
VOL	8.98M			
TXN				
SHARES/TXN				
TRAVERSE				
SHARPE 150T				

Company Description

Corning Incorporated (Corning) is a global, technology-based company that operates in five business segments: Display Technologies, Telecommunications, Environmental Technologies, Specialty Materials and Life Sciences. The Display Technologies segment manufactures glass substrates for use in liquid crystal flat panel displays. The Telecommunications segment manufactures optical fiber and cable, and hardware and equipment components for the telecommunications industry. The Environmental Technologies segment manufactures ceramic substrates and filters for automobile and diesel applications. The Specialty Materials segment manufactures products that provide more than 150 material formulations for glass, glass ceramics and fluoride crystals to meet demand for customer needs. Life Sciences segment manufactures glass and plastic consumables for pharmaceutical and scientific applications. In September 2009, the company acquired all the interest in Axygen BioScience, Inc. and its subsidiaries.

Our Comments

ALPHA-1 or -2

GOOG

GOOG		3rd QTR 2009		
DATE	**04 11 09**	Mkt cap	173.69	B
Open	543.82	Shares	316.57	M
Close	548.65	EPS	15.49	
MIN	542.66	P/E	35.45	
MAX	549.77	BETA	1.11	
Range	7.11	INST	61	%
%RANGE	0.013			
VOL	1.82M			
TXN				
SHARES/TXN				
TRAVERSE				
SHARPE 150T				

GOOG – RELATED COMPANIES

		Valuation		Profile
	Company name	Price	Mkt Cap	Employees
GOOG	Google Inc.	548.65	173.69B	19,786
MSFT	Microsoft Corporation	28.41	252.26B	93,000
YHOO	Yahoo! Inc.	15.9	22.31B	13,600
BIDU	Baidu, Inc.(ADR)	396.25	13.75B	6,387
TWX	Time Warner Inc.	31.13	36.91B	87,000
AAPL	Apple Inc.	194.18	174.89B	34,300
IBM	Intl. Business Machine...	123.1	161.70B	398,455
NWSA	News Corporation	11.94	31.21B	55,000
IACI	IAC/InterActiveCorp	19.14	2.53B	3,200
VZ	Verizon Communications...	29.31	83.26B	230,300
NOK	Nokia Corporation (ADR)	13.13	48.68B	123,347
	TECHNOLOGY	Computer services		

Company Description

Google Inc. maintains an index of websites and other online content, and makes this information freely available through its search engine to anyone with an Internet connection. The company's automated search technology helps people obtain nearly instant access to relevant information from its online index. The company generates revenue primarily by delivering online advertising. Businesses use its AdWords program to promote their products and services with targeted advertising. In August 2008, the company sold the search marketing business of Performics, a division of DoubleClick. In September 2008, Google Inc. bought Korea-based blogging software developer Tatter and Company. In September 2009, the company acquired ReCAPTCHA Inc., a spin-off of Carnegie Mellon University's Computer Science Department.

Our Comments

This we have found one of the most interesting and hardest stocks to trade. It is the highest priced stock on NASDAQ.

It appears to change character unpredictably perhaps driven by news of new developments or purchases.

It is dangerous to trade as it is capable of a 15-point drop in only a few minutes (usually recovered in about the same time).

ALPHA-2 works well with it but do watch out for drastic, steep and suddenly breaking trends.

Keep LC stop of at least 60 basis points.

GS

GS		**3rd QTR 2009**		
DATE	**07 11 09**	Mkt cap	88,309.03	M
Open	171.96	Shares	514.08	M
Close		EPS	4.62	
MIN	171.00	P/E	37.16	
MAX	173.95	BETA	1.42	
Range	2.95	INST	75	%
%RANGE	0.017			
VOL	6.28M			
TXN				
SHARES/TXN				
TRAVERSE				
SHARPE 150T				

Company Description

The Goldman Sachs Group, Inc. (Goldman Sachs) is a bank holding company and global investment banking, securities and investment management firm that provides services worldwide to corporations, financial institutions, governments and high-net-worth individuals. Its activities are divided into three segments: Investment Banking, Trading and Principal Investments, and Asset Management and Securities Services. On December 11, 2007, Credit-Based Asset Servicing and Securitization LLC, a subprime mortgage investor, completed the sale of its Litton Loan Servicing business to Goldman Sachs. In May 2008, MBF Healthcare Partners, LP and Goldman Sachs announced the acquisition of OMNI Home Care (OMNI), a provider of skilled nursing and therapy home healthcare services. MBF Healthcare Partners, LP and Goldman Sachs will share joint ownership of OMNI. In June 2008, its division, Goldman

Sachs Urban Investment Group, and Cordova, Smart & Williams, LLC announced the acquisition of H2O Plus, LLC.

Our Comments

Even with reputational damage and lawsuits galore all we care about is to make a small profit on each trade so use ALPHA-1 and possibly sell on 'money stops.'

HAL

HAL		3rd QTR 2009		
DATE	**07 11 09**	Mkt cap	27,986.84	M
Open	30.20	Shares	901.93	M
Close		EPS	1.87	
MIN	30.20	P/E	16.57	
MAX	31.58	BETA	1.52	
Range	1.38	INST	77	%
%RANGE	0.045			
VOL	15.97M			
TXN				
SHARES/TXN				
TRAVERSE				
SHARPE 150T				

Company Description

Halliburton Company provides a variety of services and products to customers in the energy industry. The company operates under two divisions: the Completion and Production, and the Drilling and Evaluation segments. Halliburton Company offers a suite of services and products to customers through its two business segments for the exploration, development and production of oil and gas. In October 2008, the company acquired Pinnacle Technologies, Inc. (Pinnacle), a provider of micro-seismic fracture mapping services and tiltmeter mapping services. In July 2008, it acquired the remaining 49% interest in WellDynamics B.V. (WellDynamics) from Shell Technology Ventures Fund 1 B.V. (STV Fund). In June 2008, Halliburton Company acquired Protech Centerform, a provider of casing centralization services. In May 2008, it acquired Knowledge Systems Inc. (KSI), a provider of combined geopressure and geomechanical analysis software and services.

Our Comments

This stock behaves very illogically much of the time as if pushed hard by internal forces.

Be careful but if you observe a trading range parameterize for a buy when it hits the support.

HESS

HESS		**3rd QTR 2009**		
DATE	**07 11 09**	Mkt cap	18,635.44	M
Open	56.34	Shares	327.05	M
Close		EPS	0.94	
MIN	56.01	P/E	60.55	
MAX	58.04	BETA	0.93	
Range	2.03	INST	78	%
%RANGE	0.036			
VOL	3.24M			
TXN				
SHARES/TXN				
TRAVERSE				
SHARPE 150T				

Company Description

Hess Corporation (Hess) is a global integrated energy company that operates in two segments: Exploration and Production (E&P) and Marketing and Refining (M&R). The E&P segment explores for, develops, produces, purchases, transports and sells crude oil and natural gas. These exploration and production activities take place principally in Algeria, Australia, Azerbaijan, Brazil, Denmark, Egypt, Equatorial Guinea, Gabon, Ghana, Indonesia, Libya, Malaysia, Norway, Russia, Thailand, the United Kingdom and the United States. The M&R segment manufactures, purchases, transports, trades and markets refined petroleum products, natural gas and electricity. As of December 31, 2008, the company owned a 50% interest in a refinery joint venture in the United States Virgin Islands, and another refining facility, terminals and retail gasoline stations located on the East Coast of the United States.

Our Comments

ALPHA-1 with the RIG and OIH caveats.

HUM

HUM		**3rd QTR 2009**		
DATE	**07 11 09**	Mkt cap	6,869.72	M
Open	39.82	Shares	169.83	M
Close		EPS	5.71	
MIN	39.44	P/E	7.09	
MAX	40.55	BETA	1.32	
Range	1.11	INST	87	%
%RANGE	0.028			
VOL	1.49M			
TXN				
SHARES/TXN				
TRAVERSE				
SHARPE 150T				

Company Description

Humana Inc. (Humana) is a health and supplemental benefits company. The company is a full-service benefits solutions company, offering an array of health and supplemental benefit plans for employer groups, government benefit programs and individuals. The company operates in two segments: Government and Commercial. The Government segment consists of beneficiaries of government benefit programs, and includes three lines of business: Medicare, Military and Medicaid. The Commercial segment consists of members enrolled in its medical and specialty products marketed to employer groups and individuals. On October 31, 2008, the company acquired PHP Companies, Inc. (doing business as Cariten Healthcare). On August 29, 2008, the company acquired Metcare Health Plans, Inc. On May 22, 2008, it acquired OSF Health Plans, Inc. On April 30, 2008, the company acquired UnitedHealth Group's Las Vegas, Nevada individual SecureHorizons Medicare Advantage health maintenance organization (HMO) business.

Our Comments

Use ALPHA-3.

ICE

ICE		3rd QTR 2009		
DATE	**07 11 09**	Mkt cap	7,765.97	M
Open	103.76	Shares	73.26	M
Close		EPS	3.8	
MIN	102.76	P/E	27.92	
MAX	106.29	BETA	1.43	
Range	3.53	INST	88	%
%RANGE	0.034			
VOL	1.09M			
TXN				
SHARES/TXN				
TRAVERSE				
SHARPE 150T				

Company Description

IntercontinentalExchange, Inc. (ICE) is an operator of regulated global futures exchanges and over-the-counter (OTC) markets. The company operates the electronic futures and OTC marketplace for trading an array of energy, soft agricultural and agricultural commodities, credit default swaps (CDS) and financial products. It offers an integrated electronic platform for side-by-side trading of energy products in both futures and OTC markets, together with clearing services. Through its electronic marketplace, it brings together buyers and sellers of derivative and physical commodities and financial contracts, and offers a range of services to support its participants' risk management needs.

Our Comments

Needs careful handling as it can be very suddenly volatile. ALPHA-1 Frequent reparameterization.

JNJ

JNJ		**3rd QTR 2009**		
DATE	**13 11 09**	Mkt cap	6.94	M
Open	27.02	Shares	134.28	M
Close		EPS	−4.92	
MIN	26.33	P/E		
MAX	27.02	BETA	1.71	
Range	0.69	INST	56.00	%
%RANGE	0.026			
VOL	2.89M			
TXN				
SHARES/TXN				
TRAVERSE				
SHARPE 150T				

Company Description

Johnson & Johnson is engaged in the research and development, manufacture and sale of a range of products in the healthcare field. Johnson & Johnson has more than 250 operating companies. The company operates in three segments. The Consumer segment includes a range of products used in the baby care, skin care, oral care, wound care and women's healthcare fields, as well as nutritional and over-the-counter pharmaceutical products. The Pharmaceutical segment includes products in the therapeutic areas, such as anti-infective, antipsychotic, cardiovascular, contraceptive, dermatology, gastrointestinal, hematology, immunology, neurology, oncology, pain management, urology and virology. The Medical Devices and Diagnostics segment includes a range of products distributed to wholesalers, hospitals and retailers. In July 2009, Johnson & Johnson completed the acquisition of Cougar Biotechnology, Inc. with an approximately 95.9% interest in Cougar Biotechnology's outstanding common stock.

Our Comments

ALPHA-2, can be very slow.

MRK

MRK		**3rd QTR 2009**		
DATE	**13 11 09**	Mkt cap	6.94	M
Open	27.02	Shares	134.28	M
Close		EPS	−4.92	
MIN	26.33	P/E		
MAX	27.02	BETA	1.71	
Range	0.69	INST	56.00	%
%RANGE	0.026			
VOL	2.89M			
TXN				
SHARES/TXN				
TRAVERSE				
SHARPE 150T				

Company Description

Merck & Co., Inc. (Merck) is a global research-driven pharmaceutical company that discovers, develops, manufactures and markets a range of products to improve human and animal health. The company's operations are principally managed on a products basis and consist of two business segments. The Pharmaceutical segment includes human health pharmaceutical products marketed either directly or through joint ventures. These products consist of therapeutic and preventive agents, sold by prescription, for the treatment of human disorders. The Vaccines and Infectious Diseases segment includes human health vaccine products marketed either directly or through a joint venture. These products consist of preventative pediatric, adolescent and adult vaccines, primarily administered at physician offices. Infectious disease products consist of therapeutic agents for the treatment of infection sold primarily to drug wholesalers and retailers, hospitals and government agencies.

Our Comments

Yet another pharma (you must have guessed by now that we consider this sector very tradable, at least at the time of writing, though *pari passu* the sector is usually in contention). ALPHA-2.

NBR

NBR		3rd QTR 2009		
DATE	**13 11 09**	Mkt cap	6,161.57	M
Open	21.78	Shares	283.29	M
Close		EPS	−0.51	
MIN	21.42	P/E		
MAX	22.99	BETA	1.66	
Range	1.57	INST	84.00	%
%RANGE	0.071			
VOL	6.60M			
TXN				
SHARES/TXN				
TRAVERSE				
SHARPE 150T				

Company Description

Nabors Industries Limited (Nabors) is a global land drilling contractor, with approximately 528 actively marketed land drilling rigs. The company conducts oil, gas and geothermal land drilling operations in the United States Lower 48 states, Alaska, Canada, South America, Mexico, the Caribbean, the Middle East, the Far East, Russia and Africa. Nabors is engaged in land well-servicing and contracting in the United States and Canada. The company markets approximately 592 land workover and well-servicing rigs in the United States, primarily in the southwestern and western United States, and actively markets approximately 171 land workover and well-servicing rigs in Canada. Nabors is a provider of offshore platform workover and drilling rigs, and actively markets 37 platform rigs, 13 jack-up units and three barge rigs in the United States and multiple international markets. The company has a 51% ownership interest in a joint venture in Saudi Arabia.

Our Comments

As RIG and OIH but try LC and ALPHA-2.

NOC

NOC		3rd QTR 2009		
DATE	**13 11 09**	Mkt cap	17,341.06	M
Open	54.69	Shares	313.75	M
Close		EPS	−3.88	
MIN	54.36	P/E		
MAX	55.50	BETA	1.01	
Range	1.14	INST	89.00	%
%RANGE	0.021			
VOL	2.27M			
TXN				
SHARES/TXN				
TRAVERSE				
SHARPE 150T				

Company Description

Northrop Grumman Corporation (Northrop Grumman) is an integrated enterprise consisting of businesses that cover the entire defense spectrum, from undersea to outer space and into cyberspace. The company is aligned into seven segments categorized into four primary businesses. The Mission Systems, Information Technology and Technical Services segments are presented as Information and Services. The Integrated Systems and Space Technology segments are presented as Aerospace. The Electronics and Ships segments are each presented as separate businesses. In January 2008, the Newport News and Ship Systems businesses were realigned into a single operating segment called Northrop Grumman Shipbuilding. In April 2008, the company announced that it had completed the sale of its Electro-Optical Systems business to L-3 Communications. In October 2008, the company acquired 3001 International, Inc. (3001).

Our Comments

Use LC.

OIH

OIH		**3rd QTR 2009**		
DATE	**07 11 09**	Mkt cap	1,886.64	M
Open	120.05	Shares	15.6	M
Close		EPS		
MIN	119.50	P/E		
MAX	122.97	BETA	1.29	
Range	3.47	INST		%
%RANGE	0.029			
VOL	6.72M			
TXN				
SHARES/TXN				
TRAVERSE				
SHARPE 150T				

Company Description

Oil Service HOLDRS Trust issues depositary receipts called Oil Service HOLDRS, representing an undivided beneficial ownership in the common stock of a group of specified companies that, among other things, provide drilling, well-site management, and related products and services for the oil service industry. The Bank of New York is the trustee. The Trust will terminate on December 31, 2041, or earlier if a termination event occurs.

Our Comments

This is one of the classic oil stocks. We trade it often with more than one algo.

ALPHA-5 algo can be magic on occasion.

Works also as lead with General Pawn.

Here we have a perfect example of how events have to be monitored. The Gulf of Mexico fiasco for BP has involved a number of subcontractors.

Steer clear for the moment.

OXY

OXY		**3rd QTR 2009**		
DATE	**07 11 09**	Mkt cap	65,647.64	M
Open	34.82	Shares	811.67	M
Close		EPS	2.98	
MIN	79.75	P/E	27.1	
MAX	81.60	BETA	1.04	
Range	1.85	INST	79	%
%RANGE	0.023			
VOL	3.46M			
TXN				
SHARES/TXN				
TRAVERSE				
SHARPE 150T				

Company Description

Occidental Petroleum Corporation (Occidental) conducts its operations through various oil and gas, chemical, midstream, marketing and other subsidiaries, and affiliates. The company operates in three business segments: the oil and gas segment, the chemical segment, and midstream, marketing and other segment. The oil and gas segment explores for, develops, produces and markets crude oil, natural gas liquids (NGLs), condensate and natural gas. The chemical segment (OxyChem) manufactures and markets basic chemicals, vinyls and performance chemicals. The midstream, marketing and other segment (midstream and marketing) gathers, treats, processes, transports, stores, trades and markets crude oil, natural gas, NGLs, condensate and carbon dioxide (CO_2) and generates and markets power. In February 2008, Occidental purchased from Plains a 50% interest in oil and gas properties in the Permian Basin and western Colorado.

Our Comments

Careful but can give good results with ALPHA-5.

PCAR

PCAR		**3rd QTR 2009**		
DATE	**07 11 09**	Mkt cap	14,075.65	M
Open	38.86	Shares	363.9	M
Close		EPS	0.49	
MIN	38.28	P/E	78.73	
MAX	39.22	BETA	1.36	
Range	0.94	INST	60	%
%RANGE	0.024			
VOL	2.17M			
TXN				
SHARES/TXN				
TRAVERSE				
SHARPE 150T				

Company Description

PACCAR Inc. operates through two segments: design, manufacture and distribution of light-, medium- and heavy-duty trucks and related aftermarket distribution of parts, and finance and leasing services provided to customers and dealers. The company's finance and leasing activities are principally related to company products and associated equipment. Other manufactured products include industrial winches. The company and its subsidiaries design and manufacture heavy-duty diesel trucks, which are marketed under the Kenworth, Peterbilt and DAF nameplates. The Truck and Other businesses include a division of the company which manufactures industrial winches in two United States plants and markets them under the Braden, Carco, and Gearmatic nameplates. In North America, Australia and 16 European countries, the company provides financing and leasing arrangements, principally for its manufactured trucks, through wholly owned finance companies operating under the PACCAR Financial trade name.

Our Comments

Approach with caution – the whipsaws may overwhelm an algo on occasion. LC.

PFE

PFE		3rd QTR 2009		
DATE	07 11 09	Mkt cap	114,465.46	M
Open	17.01	Shares	6,749.45	M
Close		EPS	1.2	
MIN	16.82	P/E	14.7	
MAX	17.02	BETA	0.71	
Range	0.20	INST	70	%
%RANGE	0.012			
VOL	36.111M			
TXN				
SHARES/TXN				
TRAVERSE				
SHARPE 150T				

Company Description

Pfizer Inc. (Pfizer) is a research-based, global pharmaceutical company. The company discovers, develops, manufactures and markets prescription medicines for humans and animals. It operates in two business segments: Pharmaceutical and Animal Health. Pfizer also operates several other businesses, including the manufacture of gelatin capsules, contract manufacturing and bulk pharmaceutical chemicals. In June 2008, Pfizer completed the acquisition of all remaining outstanding shares of common stock of Encysive Pharmaceuticals, Inc. through a merger of Pfizer's wholly owned subsidiary, Explorer Acquisition Corp., with and into Encysive. In June 2008, it also completed the acquisition of Serenex, Inc., a biotechnology company with a Heat Shock Protein 90 development portfolio. In July 2009, Pfizer bought back a 29.52% stake in its Indian arm, Pfizer Limited, increasing its stake to 70.75%. In October 2009, Pfizer Inc. acquired Wyeth.

Our Comments

This is one of the bellwethers of the pharmaceutical sector. Can be used in the 'General Pawn' algo.

RIG

RIG		3rd QTR 2009		
DATE	**07 11 09**	Mkt cap	27,425.81	M
Open	84.65	Shares	312.15	M
Close		EPS	10.11	
MIN	84.36	P/E	8.45	
MAX	86.92	BETA	0.84	
Range	2.56	INST	73	%
%RANGE	0.030			
VOL	2.95M			
TXN				
SHARES/TXN				
TRAVERSE				
SHARPE 150T				

Company Description

Transocean Ltd (Transocean), formerly Transocean Inc., is an international provider of offshore contract drilling services for oil and gas wells. As of February 3, 2009, the company owned, had partial ownership interests in or operated 136 mobile offshore drilling units. Its fleet included 39 high-specification floaters (ultra-deepwater, deepwater and harsh environment semisubmersibles, and drillships), 28 midwater floaters, 10 high-specification jack-ups, 55 standard jack-ups and four other rigs. As of February 3, 2009, the company also has 10 ultra-deepwater floaters contracted for or under construction. The company's primary business is to contract these drilling rigs, related equipment and work crews primarily on a day rate basis to drill oil and gas wells.

Our Comments

This is our number one favorite stock if it were not for the Gulf disaster. It reacts to almost all our algos and has on occasion produced spectacular results on ALPHA-5. It is a pivotal stock for the energy/oil sector. Topologically it leads the sector and is a good candidate for the 'General Pawn' algo, perhaps in tandem with OIH . . . when things return to sort of normal if they ever do.

RTN

RTN		3rd QTR 2009		
DATE	**07 11 09**	Mkt cap	18,229.63	M
Open	46.87	Shares	383.22	M
Close		EPS	462	
MIN	46.87	P/E	10.3	
MAX	47.63	BETA	0.68	
Range	0.76	INST	78	%
%RANGE	0.016			
VOL	2.35M			
TXN				
SHARES/TXN				
TRAVERSE				
SHARPE 150T				

Company Description

Raytheon Company designs, develops, manufactures, integrates, supports and provides a range of products, services and solutions for principally governmental customers in the United States and worldwide. The company operates in six business segments: Integrated Defense Systems (IDS), Intelligence and Information Systems (ibis), Missile Systems (MS), Network Centric Systems (NCS), Space and Airborne Systems (SAS) and Technical Services (TS). In April 2008, the company acquired SI Government Solutions. In July 2008, Raytheon Company acquired Telemus Solutions, Inc., a provider of information security, intelligence and technical services to defense, intelligence and other federal customers. In October 2009, the company acquired BBN Technologies.

Our Comments

This is an important and very solid stock with few surprises. We 'cut our teeth' trading this stock manually, too many years ago.

Good defense play. Use ALPHA-2. One of the rare times we did not have the LC stop in place . . . the punishment was DIRE. Unforgettable I'd say. Two weeks' hard work up in smoke.

SBUX

SBUX		**3rd QTR 2009**		
DATE	**07 11 09**	Mkt cap	15,567.55	M
Open	20.40	Shares	737.1	M
Close		EPS	0.33	
MIN	20.26	P/E	63.83	
MAX	21.29	BETA	1.35	
Range	1.03	INST	71	%
%RANGE	0.050			
VOL	28.73M			
TXN				
SHARES/TXN				
TRAVERSE				
SHARPE 150T				

Company Description

Starbucks Corporation (Starbucks) together with its subsidiaries purchases and roasts whole bean coffees and sells them, along with fresh, rich-brewed coffees, Italian-style espresso beverages, cold blended beverages, complementary food items, a selection of premium teas, and coffee-related accessories and equipment, through company-operated retail stores. Starbucks also sells coffee and tea products and licenses its trademark through other channels. Starbucks produces and sells a range of ready-to-drink beverages. The business segments of the company are United States, International and Global Consumer Products Group (CPG). The United States and International segments include company-operated retail stores and certain components of specialty operations. The CPG segment includes packaged coffee and tea sold globally through channels such as grocery stores and operates through joint ventures and licensing arrangements with consumer products business partners.

Our Comments

Having contributed substantially to the company's revenues we trade it just to get our investment back.

SHLD

SHLD		3rd QTR 2009		
DATE	**07 11 09**	Mkt cap	8,937.57	M
Open	34.82	Shares	118.81	M
Close		EPS	−0.19	
MIN	65.82	P/E		
MAX	69.22	BETA	1.42	
Range	3.40	INST	95	%
%RANGE	0.050			
VOL	1.07M			
TXN				
SHARES/TXN				
TRAVERSE				
SHARPE 150T				

Company Description

Sears Holdings Corporation (Holdings) is the parent company of Kmart Holding Corporation (Kmart) and Sears, Roebuck and Co. (Sears). The company is a broadline retailer with 2297 full-line and 1233 specialty retail stores in the United States operating through Kmart and Sears and 388 full-line and specialty retail stores in Canada operating through Sears Canada Inc. (Sears Canada), a 73%-owned subsidiary. During the fiscal year ended January 31, 2009 (fiscal 2008), Sears Holdings Corporation operated three business segments: Kmart, Sears Domestic and Sears Canada.

Our Comments

This stock was our workhorse early on and then slowly faded, perhaps due to strong institutional holdings. But it is now starting to look attractive again. ALPHA-2.

SNDK

SNDK		3rd QTR 2009		
DATE	**07 11 09**	Mkt cap	4,727.36	M
Open	20.67	Shares	227.06	M
Close		EPS	−7.98	
MIN	20.35	P/E		
MAX	20.97	BETA	1.97	
Range	0.62	INST	79	%
%RANGE	0.030			
VOL	5.69M			
TXN				
SHARES/TXN				
TRAVERSE				
SHARPE 150T				

Company Description

SanDisk Corporation (SanDisk) is engaged in designing, developing and manufacturing products and solutions in a range of form factors using the flash memory, controller and firmware technologies. The card products are used in a range of consumer electronics devices, such as mobile phones, digital cameras, gaming devices and laptop computers. The company also provides high-speed and high-capacity storage solutions, known as solid state drives (SSDs) that can be used in lieu of hard disk drives in a range of computing devices, including personal computers and enterprise servers. The company also produces universal serial bus (USB) drives, and moving picture experts group layer-3 audio (MP3) players, as well as embedded flash storage products that are used in a range of systems for the enterprise, industrial, military and other markets. In June 2008, the company completed the acquisition of MusicGremlin.

Our Comments

This company has a great opportunity, needs watching closely. In our opinion the flash drive technology will eventually completely replace computer hard disks as well as CD/DVD technology and there will be a large amount of activity with traders and institutions rebalancing their portfolios to include this stock. Use ALPHA-2 **with very frequent parameterization.**

TDW

TDW		**3rd QTR 2009**		
DATE	**07 11 09**	Mkt cap	2,259.11	M
Open	43.93	Shares	51.71	M
Close		EPS	7.18	
MIN	43.35	P/E	6.09	
MAX	45.03	BETA	1.14	
Range	1.68	INST	94	%
%RANGE	0.038			
VOL	724,050			
TXN				
SHARES/TXN				
TRAVERSE				
SHARPE 150T				

Company Description

Tidewater Inc. provides offshore supply vessels and marine support services to the offshore energy industry through the operation of offshore marine service vessels.

As of March 31, 2008, the company had a total of 430 vessels, of which ten were operated through joint ventures, 61 were stacked and 11 vessels withdrawn from service. The company provides services supporting all phases of offshore exploration, development and production, including towing and anchor handling of mobile drilling rigs and equipment; transporting supplies and personnel necessary to sustain drilling, workover and production activities; assisting in offshore construction activities; and a variety of specialized services, including pipe laying, cable laying and three-dimensional (3-D) seismic work. The company operates in two segments: United States and International.

Our Comments

As OIH and RIG.

TEVA

TEVA		**3rd QTR 2009**		
DATE	**13 11 09**	Mkt cap	46,476.92	M
Open	52.24	Shares	889.00	M
Close		EPS	0.92	
MIN	52.01	P/E	56.86	
MAX	52.45	BETA	0.20	
Range	0.44	INST	64.00	%
%RANGE	0.008			
VOL	2.68M			
TXN				
SHARES/TXN				
TRAVERSE				
SHARPE 150T				

Company Description

Teva Pharmaceutical Industries Limited (Teva) is a global pharmaceutical company that develops, produces and markets generic drugs covering all treatment categories. The company has a pharmaceutical business, whose principal products are Copaxone for multiple sclerosis and Azilect for Parkinson's disease and respiratory products. Teva's active pharmaceutical ingredient business provides vertical integration to Teva's own pharmaceutical production and sells to third-party manufacturers. The company's global operations are conducted in North America, Europe, Latin America, Asia and Israel. Teva has operations in more than 60 countries, as well as 38 finished dosage pharmaceutical manufacturing sites in 17 countries, 20 generic research and development centers operating mostly within certain manufacturing sites and 20 API manufacturing sites around the world. In January 2009, Phibro Animal

Health Corporation completed the acquisition of the Abic Animal Health business from the company.

Our Comments

Another pharma with good volatility – ALPHA-1.

UPS

UPS		**3rd QTR 2009**		
DATE	**07 11 09**	Mkt cap	54,555.92	M
Open	54.39	Shares	994.46	M
Close		EPS	1.64	
MIN	54.05	P/E	33.38	
MAX	55.03	BETA	0.80	
Range	0.98	INST	49	%
%RANGE	0.018			
VOL	3.28M			
TXN				
SHARES/TXN				
TRAVERSE				
SHARPE 150T				

Company Description

United Parcel Service, Inc. (UPS) is a package delivery company. The company delivers packages each business day for 1.8 million shipping customers to 6.1 million consignees in over 200 countries and territories. During the year ended December 31, 2008, UPS delivered an average of 15.5 million pieces per day worldwide, or a total of 3.92 billion packages. Its primary business is the time-definite delivery of packages and documents worldwide. UPS operates in three segments: US Domestic Package operations, International Package operations, and Supply Chain & Freight operations. US Domestic Package operations include the time-definite delivery of letters, documents and packages throughout the United States. International Package operations include delivery to more than 200 countries and territories worldwide. Supply Chain & Freight includes its forwarding and logistics operations, UPS Freight and other related business units.

Our Comments

Can be steady with no surprises – ALPHA-2.

WYNN

WYNN		**3rd QTR 2009**		
DATE	**07 11 09**	Mkt cap	7,353.60	M
Open	58.41	Shares	123.11	M
Close		EPS	−1.31	
MIN	58.09	P/E		
MAX	59.95	BETA	2.89	
Range	1.86	INST	65	%
%RANGE	0.032			
VOL	4.48M			
TXN				
SHARES/TXN				
TRAVERSE				
SHARPE 150T				

Company Description

Wynn Resorts, Limited (Wynn Resorts) is a developer, owner and operator of destination casino resorts. It owns and operates three destination casino resorts: Wynn Las Vegas, on the Strip in Las Vegas, Nevada, Encore at Wynn Las Vegas located adjacent to Wynn Las Vegas, and Wynn Macau, located in the Macau Special Administrative Region of the People's Republic of China (Macau). The company is also constructing Encore Wynn Macau, an expansion of its Wynn Macau resort.

Our Comments

We have traded this stock early in our research. For nearly a year we traded it manually with reasonable results. Works well with both ALPHA-1 and ALPHA-2.

Use a ten-session lookback for parameterization of ALPHA-1.

YHOO

YHOO		**3rd QTR 2009**		
DATE	**07 11 09**	Mkt cap	22,370.04	M
Open	15.89	Shares	1,403.39	M
Close		EPS	0.1	
MIN	15.76	P/E	167.67	
MAX	16.03	BETA	0.79	
Range	0.27	INST	71	%
%RANGE	0.017			
VOL	13.56M			
TXN				
SHARES/TXN				
TRAVERSE				
SHARPE 150T				

Company Description

Yahoo! Inc. (Yahoo!), together with its consolidated subsidiaries, is a global Internet brand. Together with its owned and operated online properties and services (Yahoo! Properties or Owned and Operated sites), the company also provides its advertising offerings and access to Internet users beyond Yahoo! through its distribution network of third-party entities (affiliates), who have integrated its advertising offerings into their websites, referred to as affiliate sites, or their other offerings. The company generates revenues by providing marketing services to advertisers across a majority of Yahoo! Properties and affiliate sites. The company provides services in more than 30 languages and in more than 30 countries, regions and territories, including localized versions of Yahoo! Its offerings to users on Yahoo! Properties fall into six categories: Front Doors, Communities, Search, Communications, Audience and Connected Life

Our Comments

Can be very volatile – ALPHA-2. Not our favorite.

YUM

YUM		**3rd QTR 2009**		
DATE	**07 11 09**	Mkt cap	16.48	B
Open	34.82	Shares	467.7	M
Close		EPS	2.2	
MIN	34.65	P/E	16.01	
MAX	35.34	BETA	1.05	
Range	0.69	INST	74	%
%RANGE	0.020			
VOL	5.57M			
TXN				
SHARES/TXN				
TRAVERSE				
SHARPE 150T				

Company Description

YUM! Brands, Inc. (YUM) is a quick service restaurant (QSR) with over 36 000 units in more than 110 countries and territories. Through the five concepts of KFC, Pizza Hut, Taco Bell, LJS and A&W (the Concepts), the company develops, operates, franchises and licenses a worldwide system of restaurants, which prepare, package and sell a menu of food items. The company either operates units or they are operated by independent franchisees or licensees under the terms of franchise or license agreements. In addition, the company owns non-controlling interests in unconsolidated affiliates in China who operate similarly to franchisees. YUM's business consists of three segments: United States, YUM Restaurants International (YRI) and the China Division. The China Division includes mainland China (China), Thailand and KFC Taiwan, and YRI includes the remainder of its international operations.

Our Comments

Use ALPHA-3.

CD Files List

ALGOMETRICS
ALPHA-2 (EMA PLUS) V1 AND V2
ALPHA-2 ANALYSES
ALPHA-2 PLUS RETS
ALPHA-1
ALPHA-1 BACKUP
ALPHA-2 V2 (2)
ALPHA-2
BUILDING ALPHA-1
CENTRALITY
CLUSTERING
COHORTS
COMPANY INFO TEMPLATE
CONVO
CURVES
DATA
EMA
EXCEL FUNCTIONS
EXCEL MINI SEMINAR
FCX ON ALPHA-1
HISTOGRAMS
LESHIK-CRALLE OSCILLATOR
METRICS PRICE
MOVING AVERAGES
PARAMETER SETTING
PLAIN LC OSCILLATOR
RETURNS
RETURNS(2)
STOCK DETAIL SNAPSHOT TEMPLATES
STOCK PROFILING
TRANSFORMS
VOLATILITY
WATCHLIST

Bibliography

Aldenderfer, Mark S., and Blashfield, Roger K., *Cluster Analysis* (1984) Sage Publications Inc., Newbury Park, CA.

Bass, Thomas A., *The Predictors* (1999), Henry Holt Publishing.

Bollen, B., and Inder, B. (2002) Estimating Daily Volatility in Financial Markets Utilizing Intraday Data, *Journal of Empirical Finance*, **9**(5), 551–562.

Broer, H.W., Hasselblatt, B., and Takens, F. (eds), *Handbook of Dynamical Systems* Vol. 3 (2010) North-Holland.

Broer, H.W., and Takens, F., Dynamical Systems and Chaos. *Epsilon-Uitgaven* 64 (2009); *Appl. Math. Sc.* 172, Springer-Verlag.

Casella, George and Berger, Roger L., *Statistical Inference* (2002) Wadsworth Group, Duxbury Imprint, Pacific Grove, CA.

Dacorogna, Michel M. et al., *An Introduction to High-Frequency Finance* (2001) Academic Press, San Diego, CA.

Devore, Jay L., *Probability and Statistics* (2000) Brooks/Cole, Duxbury, Pacific Grove, CA.

Duda, Richard O., Hart, Peter E., and Stork, David G., *Pattern Classification* (2001) John Wiley & Sons, Inc., NY.

Dunis, Christian et al., *Applied Quantitative Methods for Trading and Investment* (2003) John Wiley and Sons Ltd, Chichester, UK.

Everitt, Brian S., and Rabe-Hesketh, S., *The Analysis of Proximity Data* (1997) Arnold, a member of the Hodder Headline Group. Co-published in America by John Wiley & Sons Inc., NY.

Everitt, Brian S., Landau, Sabine, and Leese, Morven, *Cluster Analysis* (2001) Arnold, a member of Hodder Headline Group and co-published in the United States of America by Oxford University Press Inc., NY.

Fabozzi, Frank J. et al., *Trends in Quantitative Finance* (2006) The Research Foundation of the CFA Institute, Charlottesville, VA.

Farmer, J. Doyne, *Video on Networks and Systemic Risk* at www.Hedgehogs.net.

Farmer, J. Doyne, and Joshi, Shareen, *The Price Dynamics of Common Trading Strategies* at www.cis.upenn.edu.

Franses, Philip Hans, and van Dijk, Dick, *Non-linear Time Series Models in Empirical Finance* (2000) Cambridge University Press, Cambridge, UK.

Harman, Harry H., *Modern Factor Analysis* (1976) The University of Chicago Press, Chicago.

Hartigan, John A. *Clustering Algorithms* (1975) John Wiley and Sons Inc., NY (UMI Books on Demand custom print).

Heineman, George T., Pollice, Gary, and Selkow, Stanley, *Algorithms in a Nutshell* (2009) O'Reilly Media Inc., Sebastopol, CA.

Jajuga, K., Sokolowski, A., and Bock, H.-H. (eds) *Classification Clustering and Data Analysis* (2002) Springer-Verlag, Berlin.

Jordan, D.W., and Smith, P., *Mathematical Techniques* (1997, 2nd edn) Oxford University Press Inc., NY.

Kahneman, Daniel, *Attention and Effort* (1973) Prentice-Hall Inc., Englewood Cliffs, NJ.

Kahneman, Daniel, and Tversky, Amos, Prospect Theory: An Analysis of Decision under Risk (1979) *Econometrica* **47**(2), 263–292.

Kahneman, Daniel, Slovic, Paul, and Tversky, Amos (eds), *Judgment under Uncertainty: Heuristics and Biases* (1982) Cambridge University Press, Cambridge, UK.

Kaplan, Daniel and Glass, Leon, *Understanding Nonlinear Dynamics* (1995) Springer-Verlag, NY.

Kaufman, Leonard and Rousseeuw, Peter, J., *Finding Groups in Data* (1990) John Wiley & Sons Ltd, Chichester, UK.

Kaufman, Perry, J., *New Trading Systems and Methods* (2005, 4th edn) John Wiley & Sons Inc., Hoboken, NJ.

Kirkpatrick, Charles, D., and Dahlquist, Julie, R., *Technical Analysis* (2007) Pearson Education Inc., NJ.

Li, T.-Y., and Yorke, J.A. (1975) Conference Proceedings (American Institute of Physics, New York), No. 46, *Am. Math. Mon.* 82, 985.

Mandelbrot, Benoit, B., *Fractals and Scaling in Finance* (1997) Springer-Verlag, NY.

Mandelbrot, Benoit, B., and Hudson, Richard, L., *The Misbehavior of Markets* (2004) Basic Books, member of Perseus Books Group.

Marques de Sa, J.P. *Pattern Recognition* (2001) Springer-Verlag, Berlin.

Packard, N.H., Crutchfield, J.P., Farmer, J.D., and Shaw, R.S., Geometry from a Time Series (1980) *Phys.Rev.Lett.*, **45**(9), 712–716.

Ruelle, David, *Chance and Chaos* (1991) Princeton University Press, Princeton, NJ.

Ruelle, David, and Takens, Floris, On the nature of turbulence (1971) *Comm. Math. Phys.* 20, 167–192; 23, 343–344.

Schurmann, Jurgen, *Pattern Classification* (1996) John Wiley and Sons Inc., NY.

Ser-Huang Poon, *A Practical Guide to Forecasting financial market Volatility* (2005), John Wiley & Sons Ltd, Chichester, UK.

Sinclair, Euan, *Volatility Trading* (2008) John Wiley & Sons Inc., Hoboken, NJ.

Skiena, Steven, S., *The Algorithm Design Manual* (2008) Springer-Verlag, London.

Smith, Steve W., *Digital Signal Processing* (1997) California Technical Publishing, San Diego, CA.

Sornette, Didier, *Critical Phenomena in Natural Sciences* (2000) Springer-Verlag, Berlin.

Takayasu, H. (ed), *Empirical Science of Financial Fluctuations* (2002) Springer-Verlag, NY.

Takayasu, H. (ed), *Practical Fruits of Econophysics* (2005) Springer-Verlag, NY.

Takens, F., 'Detecting strange attractors in turbulence', in D. Rand and L.-S. Young (eds), Dynamical Systems and Turbulence, Warwick 1980, *Lecture Notes in Mathematics* 898 (1981), Springer-Verlag.

Tsay, Ruey, S., *Analysis of Financial Time Series* (2005) John Wiley and Sons Inc., Hoboken, NJ.

Wadhawan, Vinod K., 'Complexity Explained', as posted on Nirmukta.com.
Welles Wilder Jr, J., *New Concepts in Technical Trading Systems* (1978) Trend Research.
Williams, Garnett P., *Chaos Theory Tamed* (1997) Joseph Henry Press, Washington, DC.
Wishart, D. An Algorithm for Hierarchical Classifications, (1969a) *Biometrics*, 25, 165–170.
Wishart, D. Numerical Classification Method for Deriving Natural Classes, (1969b) *Nature*, 221, 97–98.
Wishart, D. A Generalised Approach to Cluster Analysis. (1971) Part of PhD Thesis, University of St Andrews.
Yoshiwara, Katherine, and Yoshiwara, Bruce, *Modelling, Functions, and Graphs* (2001) Brooks/Cole, Pacific Grove, CA.

Index